高等职业教育机电类专业"十三五"规划教材

机电专业英语

于海祥　冯艳宏　张　帆　主　编
刘　源　赵　歆　李　焱　副主编
　　　　　　李桂云　主　审

中国铁道出版社有限公司
CHINA RAILWAY PUBLISHING HOUSE CO., LTD.

内容简介

本书是高等职业院校基础英语的后续教材，主要内容包括机械技术基础、控制技术基础和机电一体化设备应用三部分，共有 13 个任务，每个任务包括机电类专业相关的英文课文、词组与短语、练习及阅读材料，为学生今后阅读相关的英文说明书及资料打下良好的英语基础。

本书适合作为高等职业院校机电一体化技术、数控技术、机械制造与自动化、工业机器人等专业的教材，也可作为相关岗位培训的参考书。

图书在版编目（CIP）数据

机电专业英语 / 于海祥，冯艳宏，张帆主编. — 北京：
中国铁道出版社，2017.6（2024.7重印）
高等职业教育机电类专业"十三五"规划教材
ISBN 978-7-113-23021-0

Ⅰ. ①机… Ⅱ. ①于… ②冯… ③张… Ⅲ. ①机电工程－英语－高等职业教育－教材 Ⅳ. ①TH

中国版本图书馆 CIP 数据核字（2017）第 092224 号

书　　名：	机电专业英语
作　　者：	于海祥　冯艳宏　张　帆

策　　划：	何红艳	编辑部电话：（010）63560043
责任编辑：	何红艳	
编辑助理：	绳　超	
封面设计：	付　巍	
封面制作：	白　雪	
责任校对：	张玉华	
责任印制：	樊启鹏	

出版发行：中国铁道出版社有限公司（100054，北京市西城区右安门西街 8 号）
网　　址：https://www.tdpress.com/51eds/
印　　刷：北京铭成印刷有限公司
版　　次：2017 年 6 月第 1 版　　2024 年 7 月第 3 次印刷
开　　本：787mm×1092mm　1/16　印张：10.5　字数：236 千
书　　号：ISBN 978-7-113-23021-0
定　　价：26.00 元

版权所有　侵权必究

凡购买铁道版图书，如有印制质量问题，请与本社教材图书营销部联系调换。电话：（010）63550836
打击盗版举报电话：（010）63549461

前 言

本书根据技能型紧缺人才培养培训工程机电一体化专业的教改方案要求，结合机电一体化专业课程教学和英语基础课程教学而编写。

本书是高等职业院校基础英语的后续教材。本书的教学目标是使学生熟悉机电一体化设备操作与维护中常见的英文，为学生今后阅读相关的英文说明书及资料，掌握机电一体化设备的操作与维护及进一步学习打下良好的英语基础。

本书包括机械技术基础、控制技术基础和机电一体化设备应用3个模块，共有13个任务。模块1主要介绍机械工程、机械制造、数控编程和公差及测量等机械技术基础知识；模块2主要介绍电气控制、液压控制、气动控制和PLC控制等控制技术基础知识；模块3主要介绍数控机床、数控电加工机床、工业机器人、自动化生产线和机电设备的安全与维护技术等机电一体化设备应用技术。

本书主要特点如下：

（1）按照学生认知规律创设功能模块，任务取材遵循"专业、实用、易学"的原则，贴近专业生产实际。

（2）任务内容选取紧密结合专业课教学内容，起到巩固专业课教学内容的作用。

（3）根据任务内容，选择恰当的图片（课文翻译部分只保留有翻译内容的图片），图文并茂，更直观，更有助于学生理解教学内容。

本书由天津中德应用技术大学于海祥，天津冶金职业技术学院冯艳宏、张帆任主编；郑州铁路职业技术学院刘源、天津中德应用技术大学赵歆、天津冶金职业技术学院李焱任副主编。具体编写分工如下：于海祥编写了模块1的任务1、任务2和模块3的任务4；冯艳宏编写了模块1的任务3、任务4，模块3的任务3、任务5；张帆编写了模块2的任务2～任务4；刘源编写了模块3的任务1；赵歆编写了模块2的任务1；李焱编写了模块3的任务2。全书由于海祥统稿。本书由天津冶金职业技术学院李桂云教授主审并对本书提出了许多宝贵意见，在此一并表示感谢！

本书适合作为高等职业院校机电一体化技术、数控技术、机械制造与自动化、工业机器人等专业的教材，也可作为相关岗位培训的参考书。

由于编者的水平有限，加之时间仓促，书中难免存在各种疏漏和不足之处，希望读者提出宝贵意见和建议。

编　者

2017年3月

CONTENTS 目 录

Module 1　Foundation of Mechanical Technology 1
　Task 1　Mechanical Engineering 1
　　Part A 1
　　　Text 1
　　　　1.1.1　Drawing 1
　　　　1.1.2　Metals and It's Properties 3
　　　　1.1.3　Heat Treatment of Metals 5
　　　　1.1.4　Machine Elements 5
　　　New Words and Phrases 7
　　　Exercises 8
　　Part B 9
　　　Reading Material：Annealing Types 9
　Task 2　Machinery Manufacturing 9
　　Part A 9
　　　Text 9
　　　　1.2.1　Metal Cutting Technology 9
　　　　1.2.2　Cutting Tool 11
　　　　1.2.3　Cutting Dosages 13
　　　New Words and Phrases 15
　　　Exercises 16
　　Part B 16
　　　Reading Material：Cutting Fluids 16
　Task 3　CNC Program 17
　　Part A 17
　　　Text 17
　　　　1.3.1　Basic CNC Program 17
　　　　1.3.2　Manual Program Code 20
　　　　1.3.3　Automatic Programming 23
　　　New Words and Phrases 25
　　　Exercises 26
　　Part B 27
　　　Reading Material：Program Configuration 27
　Task 4　Tolerance and Measure 28

Part A 28
    Text 28
        1.4.1   Tolerance 28
        1.4.2   Usages of Measure Tools 29
        1.4.3   Surface Roughness Tester 30
    New Words and Phrases 31
    Exercises 32
Part B 33
    Reading Material：Matters Needing Attention of Micrometer 33
课文翻译 34

Module 2　Foundation of Control Technology 52

Task 1　Electrical Control 52

Part A 52
    Text 52
        2.1.1   Basic Concepts of Electrical Control 52
        2.1.2   Basic Law of Electrical Control 52
        2.1.3   Reading Electrical Diagrams 55
    New Words and Phrases 59
    Exercises 60
Part B 61
    Reading Material：Electrical Control System Design 61

Task 2　Hydraulic Control 61

Part A 61
    Text 61
        2.2.1   The Work Principles of Hydraulics Transmission 61
        2.2.2   The Composition of Hydraulic Transmission System 63
        2.2.3   Characteristics and Application of Hydraulic Transmission 65
    New Words and Phrases 66
    Exercises 67
Part B 67
    Reading Material：The Operation and Maintenance of Hydraulic System 67

Task 3　Pneumatic Control 68

Part A 68
    Text 68
        2.3.1   The Working Principles of Pneumatic Transmission 68
        2.3.2   The Composition of Pneumatic Transmission System 70
        2.3.3   Characteristics and Application of Pneumatic Transmission 71

  New Words and Phrases ... 72
  Exercises ... 73
 Part B ... 73
  Reading Material：Fault Diagnosis Method for Pneumatic System ... 73
 Task 4 PLC Control ... 74
 Part A ... 74
  Text ... 74
   2.4.1 The Definition of PLC ... 74
   2.4.2 The Composition of PLC ... 74
   2.4.3 The Working Process of PLC ... 76
   2.4.4 The Applied Examples of PLC ... 76
   2.4.5 Characteristics and Development Trend of PLC ... 78
  New Words and Phrases ... 79
  Exercises ... 80
 Part B ... 80
  Reading Material：The Development of PLC ... 80
 课文翻译 ... 82
Module 3 Application of Mechanotronics Device ... 98
 Task 1 CNC Machine ... 98
 Part A ... 98
  Text ... 98
   3.1.1 The General of CNC Machine ... 98
   3.1.2 CNC Lathe ... 99
   3.1.3 CNC Machining Center ... 101
   3.1.4 Safety Notes for CNC Machine Operation ... 103
  New Words and Phrases ... 104
  Exercises ... 105
 Part B ... 106
  Reading Material：Classification of CNC Machine ... 106
 Task 2 CNC EDM Machine ... 106
 Part A ... 106
  Text ... 106
   3.2.1 CNC Sinker EDM Machine ... 106
   3.2.2 CNC WEDM Machine ... 108
  New Words and Phrases ... 111
  Exercises ... 111
 Part B ... 112

Reading Material：The Choosing of the Electrical Criteria 112
　Task 3　Robot 113
　　Part A 113
　　　Text 113
　　　　3.3.1　Robot Overview 113
　　　　3.3.2　Classification of Industrial Robot 114
　　　　3.3.3　Application of Welding Robot 117
　　　New Words and Phrases 118
　　　Exercises 118
　　Part B 119
　　　Reading Material：Development Trend of Welding Robot 119
　Task 4　Automatic Production Line 119
　　Part A 119
　　　Text 119
　　　　3.4.1　APL Overview 119
　　　　3.4.2　YL-335B APL 120
　　　　3.4.3　The Core Technology of APL 122
　　　New Words and Phrases 124
　　　Exercises 125
　　Part B 125
　　　Reading Material：Computer Integrated Manufacturing System (CIMS) 125
　Task 5　Mechanotronics Device Safety and Maintenance Technology 126
　　Part A 126
　　　Text 126
　　　　3.5.1　Safety Marks 126
　　　　3.5.2　Labor Protection 127
　　　　3.5.3　Daily Maintenance 128
　　　　3.5.4　Fault Diagnosis 128
　　　New Words and Phrases 130
　　　Exercises 131
　　Part B 131
　　　Reading Material：Color Markings 131
　　课文翻译 133
Appendix A　G-Codes and M-Codes for FANUC CNC 152
Appendix B　Abbreviations 155
References 158

Module 1　Foundation of Mechanical Technology

Task 1　Mechanical Engineering

Part A

Text

1.1.1　Drawing

1. Engineering drawing

Typical drawings in machine manufacturing are classified as part drawings and assembly drawings.

(1) Part drawings (Fig.1-1-1)

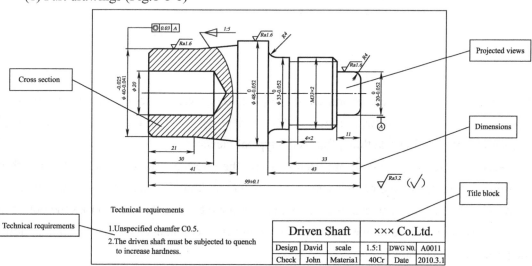

Fig.1-1-1　Part drawings

Part drawings are frequently used as instructing for manufacture and inspecting for the parts. An integrated part drawing should include a set of drawings, overall dimensions, necessary technical requirements and full contents of title block.

① A set of drawings: some representation can be given to properly show the internal and external shape of the part.

② Overall dimensions: specifying the requirements for manufacture and inspection of the integrity.

③ Necessary technical requirements: codes, symbols and notes are used to describe the essential technical requirements in the process of manufacture, inspection and assembly, such as surface roughness, tolerance, heat treatment, case treatment and the like.

④ Full contents of title block: including the part name, materials, drawing number, scale and signature of responsible individual.

(2) Assembly drawings(Fig.1-1-2)

Assembly drawings are used in explaining machines or components. In mechanical design, part drawings are usually related to the assembly drawing which indicates the working principle and structure of a machine or component. In the process of machine manufacturing, the drawings are to allow machining of the metal based on the part drawing and assembling to create a unit or a machine according to the assembly drawing.

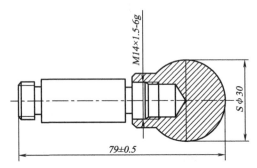

Fig.1-1-2　Assembly drawings

2. Representation of machine elements

Mechanical drawings in national standards are restricted to those which specify the view, sectional view and broken sectional view to represent the structures and shapes.

Views are projection drawings of the object. Typical views include basic views, directional views, partial views and oblique views. Basic views include upward view, front view, vertical view, right view, left view and back view, as shown in Fig.1-1-3. The number of views is always six, four and three, four is the typical views.

Sectional views (Fig.1-1-4) are used to show the internal structure of the object with a dash line. When parts have complex internal geometries, knowing the interior is as important as knowing the exterior, you can use sectioning technique to "cut sections" across the object to show internal details. We divide sectional views into full sectional views, half sectional views and partial sectional views.

Broken sectional views (Fig.1-1-5) are supposed to be sectional plane at a point somewhere on the machine element drawn only section graphics. We divide them into removed broken sectional view and superposition sectional view.

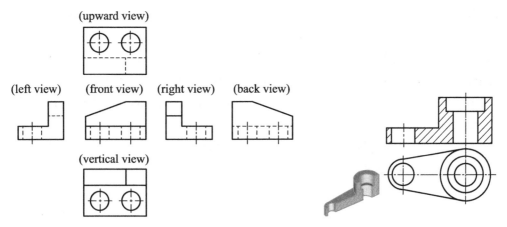

Fig.1-1-3　Basic views　　　　　Fig.1-1-4　Sectional views

For expressing the structure and size of a component distinctly, we put a component into a three-plane projection system, and acquire the three-dimensional drawing in the three projection plane. The basic principle is indicated as equal length in the front view and vertical view, equal height in the front view and left view, equal width in the vertical view and left view.

The axonometric drawing is used for complementing the outline of the body. There are two kinds of axonometric drawings: positive isometric and oblique two.

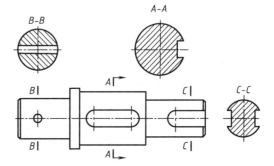

Fig.1-1-5　Broken sectional views

1.1.2　Metals and It's Properties

1. Metals

Metals are divided into two general groups: ferrous metals and nonferrous metals. The major types of ferrous metals are cast iron, carbon steels, alloy steels and tool steels.

The three primary types of cast iron are gray cast iron, white cast iron, and malleable cast iron. Gray cast iron is primarily used for cast frames, automobile engine blocks, hand-wheel and cast housings. White cast iron is hard and wear resistant and is used for parts such as train wheels. Malleable cast iron is a tough material used for tools such as pipes and wrenches. Generally, cast irons have very good compressive strength, corrosion resistance, and good machine-ability. The main disadvantage of cast iron is its natural brittleness.

The three principal types of carbon steel used in industry are low, medium, and high carbon steel. The percentage of carbon is the most important factor in determining the mechanical properties of each type of carbon steel. Low carbon contains between 0.05% and

0.30% carbon and is primarily used for parts that do not require great strength, such as chains, bolts, nuts, and pipes. Containing between 0.30% and 0.50% carbon, medium carbon steel is used for part that required great strength than is possible with low carbon steel, such as gears, crankshafts, machine parts and axles. Containing between 0.50% and 1.70% carbon, high carbon steel is used for parts that require hardness and strength, such as files, knives, drills, razors, and woodworking tools.

Alloy steels are basically carbon steels with elements added to modify of change the mechanical properties of the steel.

Tool steels are a special grade of alloy steels used for making a wide variety of tools.

Nonferrous metals are those metals whose major element is not iron. As compared to ferrous metals, the list of nonferrous metals is, of course, long and complex. The major families of nonferrous metals, such as aluminum and aluminum alloys, copper and copper alloys, magnesium and magnesium alloys, titanium and titanium alloys.

2. Properties of metals

The properties of metals are the characteristics that determine how the metal will react under varying conditions. The two principal types of properties are physical and mechanical. Physical properties are those fixed properties that are determined naturally and cannot be changer, such as weight, mass, color, and specific gravity. Mechanical properties, on the other hand, are those properties of metal that can be changed or modified to meet a particular need, such as strength, hardness, wear resistance, toughness, plasticity, and brittleness.

(1) Strength

Strength is a property of metal that allows it to resist permanent change in shape when loads are applied.There are four types or forms of strength you should know are: tensile strength, shear strength, compressive strength, and ultimate strength.

(2) Hardness

Hardness is the ability of a metal to resist indentation or penetration. Several different methods are used to measure the hardness of a metal; however, the two primary methods, or test, used by industry are the Brinell and Rockwell hardness tests.

(3) Wear resistance

Wear resistance is the ability of a metal to resist abrasion. In most cases, the harder the metal, the better it resists wear.

(4) Toughness

Toughness is the ability of a metal to resist, or absorb, sudden shocks of loads without breaking.

(5) Plasticity

Plasticity is the ability of a metal to be extensively deformed without fracture or rupture.

(6) Brittleness

Brittleness is the property of a metal that causes it to fracture rather than deform when loads are applied. Brittleness is the opposite of plasticity.

1.1.3 Heat Treatment of Metals

Heat treatment is a process of controlled heating and cooling of a metal to achieve a characteristics change in the properties. Heat treatment curve is shown as Fig.1-1-6. The five common heat-treating operations performed on steels are annealing, normalizing, hardening, tempering, and case hardening. Most nonferrous metals can be annealed, and some are harden-able by heat treatment. However, nonferrous metals are not normalized, tempered, or case hardened.

Annealing is a process used to soften metals that is generally performed on hardened parts that, for some reason, must be machined. Purposes of annealing: remove hardness; increase malleability; increase ductility; improve machine-ability; refine grain structure.

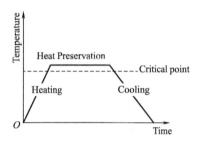

Fig.1-1-6 Heat treatment curve

Normalizing is a process used to reduce the internal stresses in a metal caused by machining or forming. Purposes of normalizing: relieve stresses; produce normal grain size and structure; place steels in the best condition for machining; lessen distortion in heat treating.

Hardening is the process of increasing the strength, hardness, and wear resistance of a metal. Purposes of hardening: increase hardness, strength and wear resistance.

Tempering relieves some of the stressed caused by rapid cooling in the hardening process. Purposes of tempering: reduce hardness to desired level; increase shock resistance and impact strength; reduce brittleness; relieve stresses caused by rapid cooling.

Case hardening is a process of producing a hard case, or shell, around a low carbon steel part by adding carbon to its surface. This process is well suited for parts that need a hard, wear-resistance surface and a tough inner core. Gears, sprockets are typical example of parts that are case hardened.

1.1.4 Machine Elements

1. Machine and parts

However simply, any machine is a combination of individual components generally referred to as machine elements or parts. Thus, if a machine is completely dismantled, a collection of simple parts such as nuts, bolts, springs, gears, cams and shafts, etc.—the building block of all machinery.

The most common example of a machine element is a gear. Gears are designed to transfer rotary motion from one shaft to another. The speed of the motion is increased or decreased by changing the size of the drive gear and the driven gear.

2. Name of different gear parts (Fig.1-1-7)

(1) Number of teeth

The total average number of gear teeth on the gear circumference is called the number of teeth, which is symbolized as Z.

(2) Addendum circle, dedendum circle

The space between the adjacent two teeth is called the keywall. The circle over the bottom of all keywalls are called dedendum circle, whose radius is symbolized as r_f. The circle over the top of all gear teeth are called addendum circle, whose radius is symbolized as r_a. The addendum circle of the external gear is bigger than its dedendum circle, the addendum circle of the internal wheel is smaller than its dedendum circle.

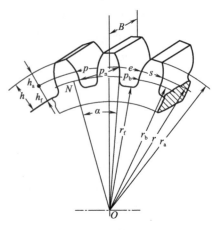

Fig.1-1-7 Name of different gear parts

(3) Reference circle

To design or manufacture, a circle is set artificially. The modulus of the circle is a standard value. Also the pressure angle is a standard value. The circle is called the reference circle.

(4) Tooth addendum, tooth dedendum, whole depth

The radial distance between the reference circle and the addendum circle is called the tooth addendum, which is symbolized as h_a.

The radial distance between the reference circle and the dedendum circle is called the tooth dedendum, which is symbolized as h_f.

The radial distance between the addendum circle and the dedendum circle is called the whole depth, which is symbolized as h.

3. Gear types and applications

There are several kinds of gears used in modern machinery. Some of those are spur gears, helical gears, gear racks, bevel gears, worm and worm wheel.

Spur gears (Fig.1-1-8) are the most widely used style of gears and are used to transmit rotary motion between parallel shafts, while maintaining uniform speed and torque. The involute tooth form, being the simplest to generate, permits high manufacturing tolerances to be attained.

Fig.1-1-8　Spur gears

Helical gears (Fig.1-1-9) are similar to spur gears with the exception that the teeth are cut at an angle to the axis of the shaft—the helix angle. The helix cut creates a wider contact area enabling higher strengths and torques to be achieved.

Bevel gears (Fig.1-1-10) are used solely to transmit rotary motion between intersecting shafts.

Fig.1-1-9　Helical gears　　　　　　　　Fig.1-1-10　Bevel gears

New Words and Phrases

roughness [rʌfnəs]	n. 粗糙，粗糙的地方
tolerance [ˈtɒlərəns]	n. 公差，限度
sectional [ˈsekʃənl]	adj. 断面的，局部的，部分（地区）的
oblique [əˈbliːk]	adj. 斜，倾斜的
gray [greɪ]	adj. 灰色的，灰白头发的
malleable [ˈmæliəbl]	adj. 可锻造的，有延展性的，韧性的
razor [ˈreɪzə(r)]	n. 剃刀，刮面刀
magnesium [mægˈniːziəm]	n. [化]镁（金属元素）
titanium [tɪˈteɪniəm]	n. [化]钛
characteristic [ˌkærəktəˈrɪstɪk]	n. 性质，特性，特征，特色
tensile [ˈtensaɪl]	adj. 拉力的，张力的，可伸展的，可拉长的
shear [ʃɪə(r)]	vi. 剪切，修剪，穿越，[力]切变
ultimate [ˈʌltɪmət]	adj. 极限的，最后的，最大的，首要的
brinell [brinel]	n. （布氏）压痕
abrasion [əˈbreɪʒn]	n. 磨损，擦伤处，磨蚀
fracture [ˈfræktʃə(r)]	n. 破裂，断裂
rupture [ˈrʌptʃə(r)]	n. 断裂，破裂

annealing [əˈniːlɪŋ]	v. 退火，退火（anneal 的现在分词）
normalizing [ˈnɔːməlaɪzɪŋ]	n. 正火
tempering [ˈtempərɪŋ]	v. 回火，钢化
malleability [ˌmælɪəˈbɪləti]	n. 有延展性，柔韧性，柔顺
distortion [dɪˈstɔːʃn]	n. 扭曲，变形，失真，畸变
dedendum [dɪˈdendəm]	n. 齿根，齿根高
addendum [əˈdendəm]	n. 附录，[机]（齿轮的）齿顶（高）
engineering drawing	工程图纸
assembly drawing	装配图
partial view	局部视图
oblique view	斜视图
cast iron	铸铁
wear resistance	耐磨性
nonferrous metal	有色金属
addendum circle	齿顶圆
dedendum circle	齿根圆
spur gears	直齿圆柱齿轮
helical gears	斜齿轮

Exercises

Ⅰ. **Match column A with column B.**

A	B
耐磨性	assembly drawings
正火	sectional views
灰铸铁	gray cast iron
剖视图	tensile strength
抗拉强度	wear resistance
齿根圆	normalizing
装配图	dedendum circle

Ⅱ. **Mark the following statements with T (true) or F (false).**

(　　) 1. Directional views include full sectional views, half sectional views and partial sectional views.

(　　) 2. The axonometric drawing is used for complementing the outline of the body.

(　　) 3. Metals are divided into three general groups: gray cast iron, white cast iron, and malleable cast iron.

(　　) 4. Brittleness is the opposite of plasticity.

( ) 5. Tempering relieves some of the stressed caused by rapid cooling in the hardening process.

( ) 6. The addendum circle of the external gear is smaller than its dedendum circle.

III. Answer the following questions briefly according to the text.

1. Which two kinds are typical drawings in machine manufacturing divided into?
2. What is the application of malleable cast iron?
3. What is the carbon content of medium carbon steel?
4. What's definition of the strength?
5. What is the function of the normalizing?

 Part B

Reading Material

Annealing Types

Annealing consists of heating steel slightly above its critical range and cooling very slowly. Annealing relieves internal stresses and strain caused by previous heat treatment, machining, or other cold working processes. The type of steel governs the temperature to which the steel is heated for the annealing process. The purpose for which annealing is being done also governs the annealing temperature.

There three types of annealing processes used in industry are full annealing, process annealing, and spheroidizing.

Full annealing is used to produce maximum softness in steel. Machinability is improved. Internal stresses are relieved. Process annealing is also called stress relieving. It is used for relieving internal stresses that have occurred during cold-working or machining processes. Spheroidizing is used to produce a special kind of grain structure that is relatively soft and machine-able. This processes generally used to improve the machine-ability.

Task 2　Machinery Manufacturing

 Part A

Text

1.2.1　Metal Cutting Technology

The six basic techniques of machining metal include turning, milling, planning, grinding, drilling and boring.

Turning is type of metal processing operation where a cutting tool is used to remove the unwanted material to produce a desired product, and is generally performed on lathe. In turning process, the rotation of spindle is the main movement, and the turning tool's move is the feed movement. Horizontal lathe is shown as Fig.1-2-1.

After lathes, milling machines are the most widely used for manufacturing applications. Vertical and horizontal milling machine is shown as Fig.1-2-2. Milling consists of machining a piece of metal by bringing it into contact with a rotating cutting tool which has multiple cutting-edges. There are many types of milling machines designed for various kinds of work. Some of the shapes produced by milling machines are extremely simple, like the slots and flat surfaces produced by circular saws. Other shapes are more complex and may consist of a variety of combinations of flat and curved surfaces depending on the shape given to the cutting-edges of the tool and on the travel path of the tool.

Fig.1-2-1　Horizontal lathe　　　　Fg.1-2-2　Vertical and horizontal milling machine

Planning metal with a machine tool is a process similar to planning wood with a hand plane. The essential difference lies in the fact that the cutting tool remains in a fixed position while the workpiece is moved back and forth beneath. Planners are usually used for processing large workpiece. A shaper (Fig.1-2-3) differs from a planer in that the workpiece is held stationary and the cutting tool travels back and forth.

Grinding machining is important processing technology in mechanical manufacturing. Grinding tool is a rotating abrasive wheel. The abrasive wheel simulates a milling cutter with a large number of miniature cutting edges. The process is often used for the final finishing process to obtain high accuracy dimensions and better surface finish of a part that has been heat-treated to make it very hard. That is because grinding can correct distortion that may have resulted from heat treatment. Grinding can be performed on plat, cylindrical and internal surface by employing specialized machining tools on grinding machine. In recent years, grinding has also found increased application in heavy-duty metal removal operations. Universal grinder is shown as Fig.1-2-4.

Module 1　Foundation of Mechanical Technology

Fig.1-2-3　Shaper　　　　　　　　　Fig.1-2-4　Universal grinder

Drilling involves producing through or blind holes in a solid metal by a cutting tool, which rotates around its axis, against the workpiece. Radial drilling machine is shown as Fig.1-2-5. Drilling operation can be carried out either by hand drill or by drilling machine. Usually, the tool rotates around its spindle while the workpiece is fixed firmly in the latter. Drilling consists of cutting a round hole by means of a rotating drill. The drill can have either one or more cutting edges and corresponding flutes which can be straight or helical. The function of flutes is to provide outlet passages for the chips generated during the drilling operation and also to allow lubricant and coolant to reach cutting edges and the surface being machined. The most common used drills in production are twist drill, center drill, gun drill and spade drill.

Boring, on the other hand, involves the finishing of a hole already drilled or cored by means of rotating, offset, single-point tool. On some boring machines, the tool is stationary and the workpiece revolves, on others, the reverse is true. Boring machine is shown as Fig.1-2-6.

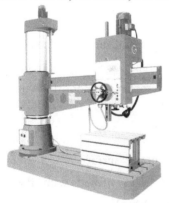

Fig.1-2-5　Radial drilling machine　　　　Fig.1-2-6　Boring machine

1.2.2　Cutting Tool

1. Cutting tool materials and property

Cutting tools properties include high hardness and the ability to retain it even at the elevated-temperatures generated during cutting. They also include toughness, abrasion resistance, and the ability to withstand high bearing pressures. A cutting material is selected to suit the cutting conditions, such as the workpiece material, cutting speed, production tare, coolants used and so on. The commonly used cutting tool materials are plain carbon steel, alloy steel, high-speed steel, cemented carbides, diamond, etc.

2. Common tools of turning

Turning operations use one cutting edge at a time. The commonly used turning tools are as follows: external turning tool [Fig.1-2-7(a)], groove tool [Fig.1-2-7(b)], thread tool [Fig.1-2-7(c)], internal turning tools [Fig.1-2-7(d)], twist drill [Fig.1-2-7(e)] and center drill [Fig.1-2-7(f)], etc.

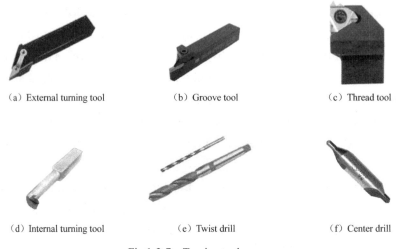

(a) External turning tool　　(b) Groove tool　　(c) Thread tool

(d) Internal turning tool　　(e) Twist drill　　(f) Center drill

Fig.1-2-7　Turning tools

External turning tool is used for turning cylinder, taper and facing surface. Groove tool is used in parting the workpiece or cut off the workpiece. Thread tools is used to cut a standard 60 degree thread. Internal turning tool is used in a boring operation. Center drill is used to provide positioning for drilling operation.

3. Common cutters of milling

The face milling cutter is used to milling of the upper surface [Fig.1-2-8(a)]; keyway cutters are used to machine contour [Fig.1-2-8(b)], holes drilled with the center drill [Fig.1-2-8(c)], twist drill [Fig.1-2-8(d)] and reamer [Fig.1-2-8(e)].

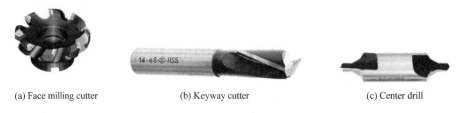

(a) Face milling cutter　　(b) Keyway cutter　　(c) Center drill

Fig.1-2-8　Milling cutters

(d) Twist drill   (e) Reamer

Fig.1-2-8 Milling cutters (continued)

4. Geometry of cutting tool

The cutting part of cutting tool includes a face for passing chips and a flank directed to the workpiece. The intersecting face and flank form a cutting edge. The tool performance depends on its material and angles which mainly include: nose angle, rake angle, relief angle and cutting edge angle.

Rake angle decides the tartness degree of tool, the larger of the rake angle, the more tartness. Rake angles can be positive, negative, or zero. Its value usually varies between 0° and 15°, whereas the back rake angle is usually taken as 0°.

Relief angles serve to eliminate rubbing between the workpiece and the end flank. The degree of relief angle has important effect on surface quality of the workpiece. At the same time, relief angle affects the intensity of tool edge. It can also affect the tartness degree of tool. Usually, the values of each of these angles range between 4° and 6° (rough machining) or 8° and 12° (finish machining).

1.2.3 Cutting Dosages

1. Cutting speed

Cutting speed for milling is the speed at the outside edge of the milling cutter as it is rotating. A milling cutter must spin. The rate at which the cutting tool rotates is called the spindle speed, measured in r/min (revolutions per minute).

Each cutter has its own spindle speed, depending on the type of material being cut and the size (diameter) of the cutter. When cutting the same type of material, the smaller the cutter, the faster it must rotate.

For lathe operation, the cutting speed is defined as the rate at which a point on the circumference of the workpiece passes the cutting tool. It is the number of feet traveled in the circumferential direction by a given point on the surface of the workpiece per minute. The relationship between the cutting speed and spindle speed can be given by the following equation:

$$\text{Cutting speed} = \pi D n / 1000$$

Where: D—diameter of the workpiece, mm;

n—spindle speed, r/min.

Every material has an ideal cutting speed. This is the optimum speed at which the material can be cut safely, in order to obtain a good quality of finish. The cutting speed is dependant primarily upon the material being machined as well as the material of the cutting tool and can be obtained from handbooks, provided by cutting tool manufacturers. There are also other variables that affect the optimal value of the cutting speed. These include the tool geometry, the type of lubricant or coolant, the feed rate, and the depth of cut.

2. Feed rate

The feed rate is that the distance the tool advances into the workpiece per revolution of the workpiece. The selection of a suitable feed rate depends upon many factors, such as the required surface finish, the depth of cut and the geometry of the tool used. Finer feed rate produces better surface finish, whereas higher feed rate reduces the machining time. Therefore, it is generally recommended to use higher feed rate for roughing operations and finer feed rate for finishing operations. Recommended values for feed rate, which can be taken as guidelines, are found in handbooks provided by cutting tool manufacturers.

Feed rate is usually measured in inches per minute. For milling operation it is the revolution per minute times the number of teeth in the cutter. Due to variations in cutter sizes, number of teeth and revolutions per minute, all feed rate should be calculated from feed per tooth.

The feed rate per tooth must be converted into feed rate per minute before you can make the feed rate setting on the machine. The formula for converting feed rate is as follows:

$$\text{feed rate (in/min)} = \text{spindle speed} \times \text{chip load} \times \text{teeth}$$

It must also be mentioned that using a chip load that is too small will cause excessive tool wear, so don't just set the feed rate low and think this is correct.

3. Cut depth

Depth of cut is defined as the distance that the cutting tools is plunged into the workpiece. It is typically measured in millimeters. For turning operation, the depth of cut can be calculated by the following equation:

$$a_p = (d_w - d_m)/2$$

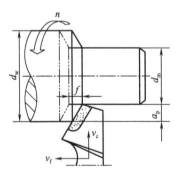

Fig.1-2-9 Depth of cut

Where: a_p—depth of cut(Fig.1-2-9);

d_w—the diameter of the new surface(Fig.1-2-9);

d_m—the diameter of the finished surface(Fig.1-2-9).

To determine the depth of cut we must first select the proper cutting tool, the proper machine and a suitable setup. The depth of cut directly influences the tool life. If the depth of cut is too large, the insert may be overloaded, causing immediate breakage. The handbooks provide the recommended ranges for depths of cut (a_p) for each insert. If the depth of cut is too small, the resulting side forces will not be sufficient to properly deflect the tool. Vibration and instability may occur. In the finishing operation, it is important to select a small depth of cut and a small corner radius.

If your CNC machine is used with incorrect spindle speed, feed rate and depth of cut, your work may be machined with a poor surface finish or the workpiece or cutter could be damaged. There are many factors that would affect these values, including:

- The condition of the machine.
- The type of material being machined.
- The clamping method used to the stock on the CNC machine.
- The type of cutting tool used.
- The diameter of the cutting tool.
- Material type of the cutting tool.

New Words and Phrases

grinding [ˈgraɪndɪŋ]	v. 磨碎，嚼碎（grind 的现在分词）
boring [ˈbɔːrɪŋ]	n. 钻孔，钻屑
abrasive [əˈbreɪsɪv]	adj. 有磨蚀作用的，摩擦的，粗糙的
abrasion [əˈbreɪʒn]	n. 磨损，擦伤处
resistance [rɪˈzɪstəns]	n. 阻力，抵抗，抗力
cemented carbides	烧结碳化物，硬质合金
diamond [ˈdaɪəmənd]	n. 钻石，金刚石
groove [gruːv]	n. 沟，槽
keyway [ˈkiːweɪ]	n. 键沟
flank [flæŋk]	n. 侧面，侧腹，侧边
thermal [ˈθɜːml]	adj. 热的，保热的，温热的
consumption [kənˈsʌmpʃn]	n. 消费，耗尽
chip [tʃɪp]	n. 碎片 vt. 刻，削成，从……上削下一小片
synthetic [sɪnˈθetɪk]	adj. 合成的，人造的 n. 合成物，合成纤维，合成剂
emulsion [ɪˈmʌlʃnz]	n. 乳剂，乳胶
rake angle	倾角，前角，翘角
relief angle	后角
cutting edge	切削刃
plain carbon steel	碳素钢
groove tool	切槽刀
cutting dosage	切削用量
feed rate	进给量
recommended value	推荐值

cutting fluid 切削液
synthetic oil 合成油

Exercises

Ⅰ. Match column A with column B.

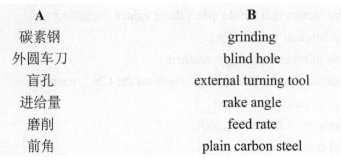

A	B
碳素钢	grinding
外圆车刀	blind hole
盲孔	external turning tool
进给量	rake angle
磨削	feed rate
前角	plain carbon steel

Ⅱ. Mark the following statements with T (true) or F (false).

(   ) 1. In turning process, the rotation of spindle is the feed movement, and the turning tool's move is the main movement.

(   ) 2. Groove tool is used in parting the workpiece or cut off the workpiece.

(   ) 3. The twist drill is the most common type of drill and used for rough drilling operation.

(   ) 4. Relief angle decides the tartness degree of tool, the larger of the relief angle, the more tartness.

(   ) 5. The feed rate per tooth is too small will cause excessive tool wear so don't just set the feed rate low and think this is correct.

Ⅲ. Answer the following questions briefly according to the text.

1. How many kinds are there in techniques of machining metal?
2. What function is relief angle?
3. What are the factors that affect the cutting speed?
4. How to select the feed rate for roughing operation?
5. What is the definition of cut depth?

 Part B

Reading Material

Cutting Fluids

1. Functions of cutting fluids

① Reduce friction and wear, thus improving tool life and the surface finish of the workpiece.

② Cool the cutting zone, thus improving tool life and reducing the temperature and thermal distortion of the workpiece.

③ Reduce forces and energy consumption.

④ Flush away the chips from the cutting zone, and thus prevent the chips from interfering with the cutting process.

⑤ Protect the machined surface.

2. Types of cutting fluids

(1) Synthetic oils

Synthetic oils typically are used for low-speed operations where temperature rise is not significant.

(2) Emulsions

Emulsions are a mixture of oil and water and additives, generally are used for high-speed operations because temperature rise is significant. The presence of water makes emulsions very effective coolants.

Task 3　CNC Program

 Part A

Text

1.3.1　Basic CNC Program

1. Coordinate systems (Fig.1-3-1)

The machine manufacturer sets a machine zero point for each machine. The machine zero point is the origin of the machine coordinate system. When the operator wants to determine the position of the workpiece on the machine, he must set the workpiece coordinate systems. If the dimension of a workpiece is too big, the user can set another coordinate system in a local area of the workpiece, this is a local coordinate system.

The coordinates in a coordinate system represent the tool position. Coordinates axes (Fig.1-3-2) are used to specify the coordinates. Three main axes are referred to as the X, Y and Z axes. The Z axis is perpendicular to both the X and Y axes in order to create a right-hand coordinate system. A positive motion in the direction moves the cutting tool away from the workpiece.

2. Program zero

The origin point for each axis is commonly called the program zero point, also called work zero, part origin or zero point. This point is usually the location on the print from which all (or the most) dimensions are taken. This is commonly done with a G92 (or G50) command at the beginning of the program.

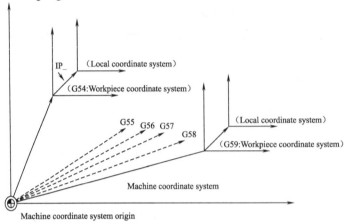

Fig.1-3-1   Coordinate systems

The placement of program zero is determined by the programmer. Program zero could be placed anywhere. The wise selection of a program zero point will make programming much easier. You should always make your program zero point a location on the datum surfaces of your workpiece. There is a program zero point for each axis.

The program zero of a lathe is generally set in the midpoint of the workpiece's right surface, see Fig.1-3-3.

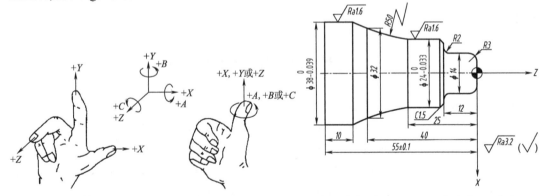

Fig.1-3-2   Coordinates axes   Fig. 1-3-3   Program zero of a lathe workpiece

3. Absolute and incremental coordinate (Fig.1-3-4)

If the tool moves to a target point, the coordinate value of it is determined on the basis of zero point of the workpiece coordinate system. This value is the absolute coordinate value. If the tool moves from the present point to a target point, the coordinate value of the target point is determined on the basis of the previous tool's coordinate value. This value is the incremental coordinate value.

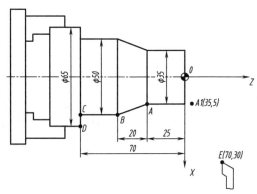

Absolute			Incremental		
Point	X	Z	Point	U	W
E	70	30	E	0	0
A1	35	5	A1	−35	−25
A	35	−25	A	0	−30
B	50	−45	B	15	−20
C	50	−70	C	0	−25
D	65	−70	D	15	0

Fig.1-3-4 Movement in both the absolute and incremental coordinate

4. The G, F, S, M and T-Codes

G-codes are also called preparatory codes or words. A preparatory code is designated by the address G followed by one or two digits to specify the mode in which a CNC machine moves along its programmed axes. G-codes are usually classified into two types. The modal G-codes specification will remain in effect for all subsequent blocks unless replaced by another G-code of same group. The non-modal G-codes specification will only affect the block that it appears.

The F-code (Fig.1-3-5) controls the speed of the cutting feed. It can be expressed as feed per minute or feed per revolution. An F specification is modal and remains in effect in a program for all subsequent tool movements. Feed speed can be changed frequently in a program, as needed.

The S-code (Fig.1-3-6) controls the spindle speed. The selected speed value right follows the S address. The S command should be given together with the spindle revolution command (M03 or M04).

M-codes are called the miscellaneous words and are used to control miscellaneous function of the machine. Such functions include turn the spindle on/off, start/stop the machine, turn on/off the coolant, change the tool, and rewind the program tape.

T-code(Fig.1-3-7) is used to specify the tool number. It is used only for an automatic tool changer machine.

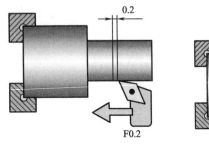
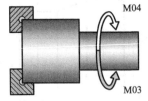
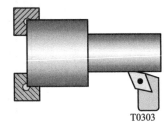

Fig.1-3-5 F-code   Fig.1-3-6 S-code   Fig.1-3-7 T-code

1.3.2 Manual Program Code

1. Basic program codes

(1) G00 (rapid positioning)

The G00 command moves a tool to the position in the workpiece system specified with an absolute or an incremental command at a rapid traverse rate. In the absolute command, coordinate value of the end point is programmed. In the incremental command, the distance the tool move is programmed.

(2) G01 (linear interpolation)

The G01 command moves a tool along a line to the specified position at the feed rate specified in F-code. The feed rate specified in F-code is effective until a new value is specified. It need not be specified for each block.

(3) G02/G03 (circular interpolation)

The command will move a tool along a circular arc. The arc center is specified by addresses I, J and K for the X, Y and Z axes, respectively. The numerical value following I, J or K, however, is a vector component in which the arc center is seen from the start point.

(4) G17/G18/G19 (plane selection)

We must select plane before some program codes are programmed. For example: G17: specification of tool path on XY plane; G18: specification of tool path on ZX plane; G19: specification of tool path on YZ plane.

2. Tool compensation codes

(1) Tool length compensation (Fig.1-3-8)

You will instate tool length compensation in each tool's first Z axis approach movement to the workpiece. This instating command will include a G43 or G44 code and an H word to invoke the related tool offset. This instating code must also contain a Z axis positioning movement. Once instated, tool length compensation will remain in effect until the next G43 or G44 code in the first Z axis command of the next tool (tool length compensation is modal).

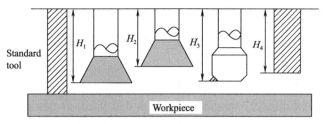

Fig.1-3-8　Tool length compensation

G49 code can cancel tool length compensation.

(2) Tool radius compensation (Fig.1-3-9)

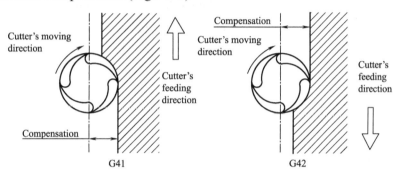

Fig.1-3-9　G41 and G42

Most controls use three G-codes with tool radius compensation. G41 is used to instate a cutter left condition (climb milling). G42 is used to instate a cutter right condition (conventional milling). G40 is used to cancel cutter radius compensation. Additionally, many controls use a D word to specify the offset number used with tool radius compensation.

To determine whether to use G41 or G42, simply look along the cutter is moving during machining. If the cutter is on the left side of the surface being machined, use G41. If right, use G42. Once tool radius compensation is properly instated, the cutter will be kept on the left side or right side of all surfaces until the G40 code to cancel tool radius compensation.

3. Multiple repetitive cycle codes

There are several types of predefined canned cycle codes that make programming easier.

(1) G71 (stock removal in turning)

This canned cycle will rough out material on a part given the finished part shape. All that a programmer needs to do is define the shape of a part by programming the finished tool path and then submitting the path definition to G71 by means of a PQ block designation.

(2) G73 (pattern repeating in turning)

This canned cycle permits cutting a fixed pattern repeatedly, with a pattern being displaced bit by bit. By this cutting cycle, it is possible to efficiently cut work whose rough shape has already been made by a rough machining, forging or casting method, etc.

(3) G70 (finishing cycle in turning)

After rough cutting by G71 or G73, the G70 finishing cycle can be used to finish cut paths.

(4) G92 (simple threading cycle) (Fig.1-3-10)

The main benefit of the G92 threading cycle is that it eliminates such repetitive data and makes program easer to edit. In the G92 threading cycle application, input of each thread pass diameter is important.

(5) G81 (drilling cycle)

The programmer can choose G81 code when drilling shallow holes.

G81 code is programmed by entering in a block of information: the X and Y coordinates, the Z axis reference plane (R) and the final Z axis depth. The code will cause the following sequence of operations to occur (see Fig.1-3-11):

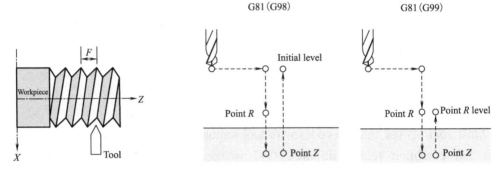

Fig.1-3-10　Threading cycle　　　　　　　Fig.1-3-11　G81

① Rapidly move to the X and Y coordinate of the hole position.

② Rapidly move to Z axis reference plane (R).

③ Feed to the Z axis final depth.

④ Rapidly back to either the Z axis initial position or the Z axis reference plane (R).

(6) G80 (canned cycle for drilling cancel)

Canned cycle for drilling is cancelled to perform normal operation. Point R and Point Z that are specified in G81 are cleared. Other drilling data is also cleared.

4. Subprogram

If a program contains a fixed sequence or frequently repeated pattern, such a sequence or pattern can be stored as a subprogram in memory to simplify the program. The subprogram can be called from the main program, and the called subprogram can also call another subprogram. Subprogram call is shown as Fig.1-3-12.

A single call program code(M98) can repeatedly call a subprogram up to 9,999 times. When the main program calls a subprogram, it is regarded as a one-level subprogram call. Thus, subprogram calls can be nested up to four levels.

When M99 code is executed in a subprogram, the subprogram is terminated and control returns to the block after the calling block or the block with the sequence number specified by P.

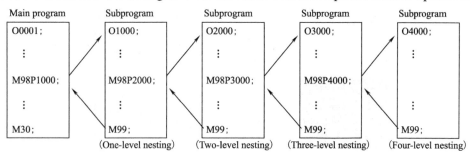

Fig.1-3-12 Subprogram call

1.3.3 Automatic Programming

There are many types of automatic programming software such as MasterCAM , UG , Pro/E, etc.

MasterCAM X is most widely used software for CNC automatic programming. The main functions are created the geometry modeling, create the tool path, verify the cutting process, create the G-codes.

Here we will study MasterCAM X for CNC automatic programming.

1. Create the geometry modeling

You can create the geometry in one of two ways:

① By using the graphical design interface provided by MasterCAM X.

② By making the design in CAD software, e.g. UG , Pro/E , SolidWorks then saving it in a format that MasterCAM X can import.

2. Creating the tool path

(1) Select the machine type (Fig.1-3-13)

Click "Machine Type"→"Mill", select the machine type.

(2) Look for an existing tool that you may want to use (Fig.1-3-14, Fig.1-3-15)

Click "Toolpaths"→"Tool Manager", in the "Tool Manager" dialog box, all available tools are displayed. You can search for a particular type by scrolling down the list or by using the Filter button and entering the search information.

(3) Create the tool path (Fig.1-3-16)

① Click "Toolpaths"→"Surface Rough"→"Rough Parallel Toolpath", dialog box will pop up, select "Boss", click ✓ button.

② Click on the surfaces that you wish to machine on the part. Make sure that all the surfaces been selected then click ✓ button.

③ Setting surface rough parallel parameter

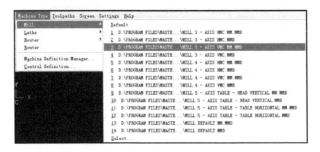

Fig.1-3-13   Select the machine type

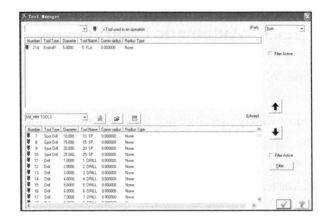

Fig.1-3-14   "Tool Manager" dialog box

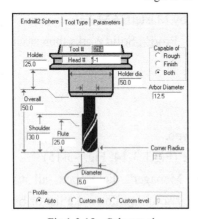

Fig.1-3-15   Select tool

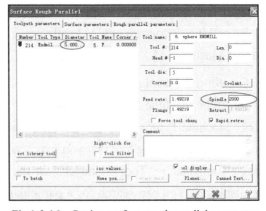

Fig.1-3-16   Setting surface rough parallel parameter

3. Verify the cutting process (Fig.1-3-17, Fig.1-3-18)

Before a part is machined, the CAM model needs to be verified that the part program is correct. The purposes of verification are:

① To detect geometric error of the cutter path.

② To detect potential tool interference.

③ To detect erroneous cutting conditions.

To begin verification, select one more operations in the "Toolpath Manger list", then click the verify button, use the control buttons located at the top of the "Verify" dialog box to running simulation.

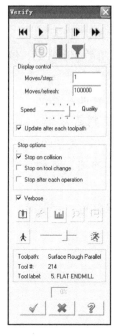

Fig.1-3-17 "Verify" dialog box

Fig.1-3-18 The result of verification

4. Create the G-codes (Fig.1-3-19)

Different CNC machines use slightly different versions of G-codes. The conversion of the machining data to the G-codes specific for a particular CNC machine is called post-processing. The format of the G-code is stored in different post-processing files and the system will use whichever post-processing format you select.

From the left panel in the "Toolpaths tab", select the G1 icon, the "Post Processing" dialog box pops up, click button, input the file name. Once the G-codes are generated, we will allow you to see the G-codes file and modify if required.

```
0001 %
0002 O0000 (                    1)
0003 (DATE=DD-MM-YY - 25-12-09 TIME=HH:MM - 08:30)
0004 (MCX FILE - C:\DOCUMENTS AND SETTINGS\JINLONG\桌面\MASTERCAM资料\T-1.MCX)
0005 (NC FILE - C:\DOCUMENTS AND SETTINGS\JINLONG\桌面\                    1.NC)
0006 (MATERIAL - ALUMINUM MM - 2024)
0007 ( T214 |   5. FLAT ENDMILL | H0 )
0008 ( T1 |   5. FLAT ENDMILL | H0 )
0009 N100 G21
0010 N102 G0 G17 G40 G49 G80 G90
0011 N104 T214 M6
0012 N106 G0 G90 G54 X-100. Y-50. S0 M5
0013 N108 G43 H0 Z42.794
0014 N110 Z40.794
0015 N112 G1 Z37.794 F1.8
0016 N114 X100.
0017 N116 G0 Z38.794
0018 N118 Z40.794
0019 N120 Z42.794
0020 N122 X-100. Y-45.
0021 N124 Z40.794
0022 N126 G1 Z37.794
0023 N128 X100.
0024 N130 G0 Z38.794
0025 N132 Z40.794
0026 N134 Z42.794
0027 N136 X-100. Y-40.
0028 N138 Z40.794
0029 N140 G1 Z37.794
0030 N142 X100.
0031 N144 G0 Z38.794
0032 N146 Z40.794
0033 N148 Z42.794
0034 N150 X-100. Y-35.
0035 N152 Z40.794
0036 N154 G1 Z37.794
0037 N156 X100.
0038 N158 G0 Z38.794
0039 N160 Z40.794
0040 N162 Z42.794
```

Fig.1-3-19 G-codes

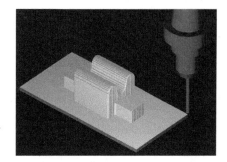

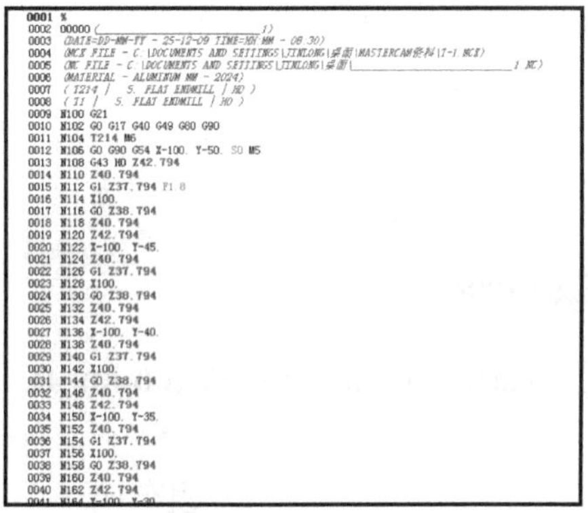

New Words and Phrases

absolute [ˈæbsəluːt] *adj.* 绝对的，完全的 *n.* 绝对，绝对事物

incremental [ˌɪŋkrəˈmentl]	adj. 增加的
subprogram [ˈsʌbprəʊɡræm]	n. 子程序，辅程序
subsequent [ˈsʌbsɪkwənt]	adj. 随后的，后来的，附随的
modal [ˈməʊdl]	adj. 模式的，情态的，形式的，语气的
compensation [ˌkɒmpenˈseɪʃn]	n. 补偿，赔偿，修正，补救办法
offset [ˈɒfset]	vt. 抵消，补偿　　adj. 分支的，偏（离中）心的
climb [klaɪm]	v. 爬，攀登，上升，登山
conventional [kənˈvenʃənl]	adj. 传统的，习用的，平常的，依照惯例的，约定的
repetitive [rɪˈpetətɪv]	adj. 重复的，啰嗦的
pattern [ˈpætn]	n. 模式，图案，样品，典范
verify [ˈverɪfaɪ]	vt. 核实，证明，判定
interface [ˈɪntəfeɪs]	n. 界面，[计]接口，交界面
interference [ˌɪntəˈfɪərəns]	n. 干涉，干扰，冲突，抵触
simulation [ˌsɪmjuˈleɪʃn]	n. 模仿，模拟
slightly [ˈslaɪtli]	adv. 轻微地，轻轻地，细长地
post-processing [ˈpəʊstprˈəsesɪŋ]	后处理
program zero point	程序零点
absolute coordinate	绝对坐标
incremental coordinate	增量坐标
rapid positioning	快速定位
linear interpolation	直线插补
tool compensation	刀具补偿
stock removal in turning	粗车循环
pattern repeating in turning	仿型粗车循环
automatic programming software	自动编程软件
tool path	刀具路径

Exercises

Ⅰ. Match column A with column B.

A	B
仿型粗车循环	incremental coordinate
增量坐标	subprogram
几何模型	tool compensation
刀具补偿	pattern repeating in turning
子程序	drilling cycle
钻孔循环	geometry modeling

II. **Mark the following statements with T** (true) **or F** (false).

(   ) 1. The operator sets a machine zero point for each machine.

(   ) 2. A positive motion in the Z direction moves the cutting tool away from the workpiece.

(   ) 3. Generally speaking, there are three types of program: the program number, the program content and the program end.

(   ) 4. Many controls use a D word to specify the offset number used with tool radius compensation.

(   ) 5. A single call program code (M98) can only call a subprogram once.

(   ) 6. In MasterCAM you can't see the G-codes file.

III. **Answer the following questions briefly according to the text.**

1. Who determines the program zero?

2. What does the program number usually start with?

3. Which command moves a tool to the position in the workpiece system specified with an absolute or an incremental command at a rapid traverse rates.

4. When had we better use subprogram?

 Part B

Reading Material

Program Configuration

Generally speaking, there are two types of program: the main program and the subprogram. The main program contains a series of commands for machining workpieces. The subprogram can be called by the main program or another subprogram. Either the main program or a subprogram contains three parts: the program number, the program content and the program end.

Programs are stored in the MCU memory by program number. Most machines can store several different programs at a time. The program number usually starts with "O" (or %). The number after "O" (or %) is the program number. Program numbers ranges from 01 to 9,999. The content of the program is the core of the whole program. It is made up of many blocks. These blocks control the movements that the CNC machine is to execute. The M02 or M30 command is used to stop the whole program.

A CNC program consists of one or more blocks of commands. A block is a complete line of information to the CNC machine. It consists of one word or an arrangement of words. Blocks may vary in length. Thus, a programmer needs to master all the words required to execute a particular machine function.

Task 4　Tolerance and Measure

Part A

Text

1.4.1　Tolerance

In dimensioning a drawing, the numbers placed in the dimension lines represent dimensions that are only approximate and do not represent any degrees of accuracy unless so stated by the designer. The numbers are termed as nominal size. The nominal size of a component dimension is arrived at as a convenient size based on the design process. However, it is almost impossible to produce any component to the exact dimension through any of the known manufacturing processes. It is therefore customary in engineering practice to allow a permissible deviation from the nominal size, which is termed as tolerance. Tolerance on a dimension can also specify the degree of accuracy. For example, a shaft might have a nominal size of 33.5 mm, if a variation of ±0.05 mm could be permitted, the dimension would be stated (33.5±0.05)mm.

In engineering when a product is designed it consists of a number of parts and these parts mate with each other in some form. In the assembly it is important to consider the type of mating or fit between two parts which will actually define the way the parts are to behave during the working of the assembly. Take for example a shaft and hole, which will have to fit together. In the simplest case if the dimension of the shaft is lower than the dimension of the hole, then there will be clearance. Such a fit is termed clearance fit. Alternatively, if the dimension of the shaft is more than that of the hole, then it is termed interference fit.

Generally speaking, the cost of a part goes up as the tolerance is decreased. If a part has several or more surfaces to be machined, the cost can be excessive when little deviation is allowed form the nominal size.

Allowance, which is sometimes confused with tolerance, has an altogether different meaning. It is the minimum clearance space intended between mating parts and represents the condition of the tightest permissible fit. If a shaft, size $\phi 1.498_{-0.003}^{0}$ is to fit a hole of size $\phi 1.500_{0}^{+0.003}$, the minimum size hole is 1.500 and the maximum size shaft is 1.498. Thus the allowance is 0.002 and the maximum clearance is 0.008 as based on the minimum shaft size and maximum hole dimension.

There are two types of tolerance: dimensional tolerance and geometric tolerance.

Dimensional tolerance can be symmetrical, such as 50±0.1, or asymmetrical, such as $\phi 50_{-0.045}^{-0.030}$, the dimension 50 is basic size. And the upper deviation is -0.030, the lower deviation is -0.045.Therefore the maximum limit size is 49.970. The minimum size is 49.955.

Geometric tolerance is used to specify the features of shape and position. Things like: straightness, flatness, circularity, cylindricity, perpendicularity, etc. Geometric characteristic symbols are shown in Table 1-4-1.

Table 1-4-1 Geometric characteristic symbols

Type	Symbol	Type	Symbol
straightness	——	angularity	∠
flatness	▱	parallelism	//
circularity(roundness)	○	position	⊕
cylindricity	⌭	concentricity	◎
profile of line	⌒	symmetry	≡
profile of surface	⌓	circular run out	↗
perpendicularity	⊥	total run out	⌰

1.4.2 Usages of Measure Tools

1. Usage of a vernier caliper

The vernier caliper (Fig.1-4-1) is a measuring instrument consisting of an L-shaped frame with a linear scale along its longer arm (called main scale) and an L-shaped sliding attachment with a vernier, used to read directly the dimension of an object represented by the separation between the inner or outer edges of the two shorter arms (called caliper jaw).

The vernier caliper provides the three basic functions of inner, outer and depth gauge.

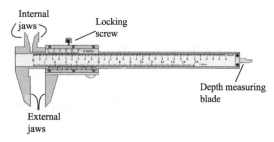

Fig.1-4-1 Vernier caliper

The vernier caliper is first set over the object to be measured with the caliper jaws shut so that it is firm but not tight. The scale is then read by first taking note of where the zero mark on the vernier scale falls on the main scale. This is the number of complete divisions on the main scale. This is the whole number that should be noted. The fraction or decimal is then read from the vernier scale. This number is taken as the line on the vernier scale that aligns with any line on the main scale.

Regardless of what type of vernier caliper you use, be sure to take the following precautions to avoid damaging the vernier caliper:

① Wash your hands before you handle the vernier caliper to remove dirt and oils that might damage the vernier caliper.

② Wipe the caliper vernier components clean both before and after you use the vernier caliper.

③ Do NOT drop the vernier caliper, which may damage or destroy the vernier caliper.

2. Usage of an outside micrometer

Outside micrometer usually called micrometer for short. It is an even more rigid measuring instrument than vernier caliper. Fig.1-4-2 shows a common outside micrometers, which range is 0~25 mm. It usually consists of sleeve, thimble, anvil, spindle, ratchet, etc.

There is a horizontal line on the micrometer sleeve. Each side of this line, there is a list of calibration line, the space between them is 1mm. The above calibration line just lies on the middle of the two below adjacent calibrations. The thimble is divided into 50 equal parts. It is whirligig. One complete rotation of the thimble makes the spindle go ahead or back one pitch of screws — 0.5 mm.

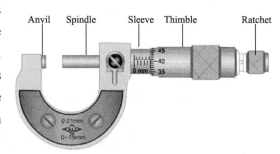

Fig.1-4-2 Outside micrometer

To read a micrometer, you place the material to be measured between the anvil and spindle, and then turn the ratchet until the spindle closes down and stops moving. First, take the thimble's end surface as the directrix, you read the markings on the sleeve (read the integer part only); then take the level line on the sleeve as the directrix, read the making on the thimble. The sum of these two parts is the result, the former is integer part and the latter is decimal part.

1.4.3 Surface Roughness Tester

1. Outline of the surface roughness tester

The SJ-201P is a shop-floor type surface roughness measuring instrument, which traces the surfaces of various machine workpiece, calculates their surface roughness based on standards roughness specimen and displays the results.

2. SJ-201P operation panel (Fig.1-4-3)

3. SJ-201P measurement operation

The procedures of the measurement operation of using the SJ-201P are as follows:

① Press the [POWER/DATA] key to turn the power on.

② Calibration. The process of calibration involves the measurement of a standard roughness specimen and the adjustment of the difference between the measured value and the standard roughness specimen. The following are the calibration procedures.

a. Press the [CAL/STD/RANGE] key in the measurement mode and the calibration value is displayed.

b. If the displayed value is different form that marked on the precision roughness specimen, modify the calibration value. If the calibration value does not require modifiication, press the [n/ENT] key. The calibration value has now been set.

c. Setup the roughness specimen.

d. Press the [START/STOP] key. A calibration measurement with the precision roughness specimen is performed, and "-----" is displayed during the calibration measurement. When the calibration measurement has been completed, the measured value will be displayed. Press the [n/ENT] key to complete the entire calibration.

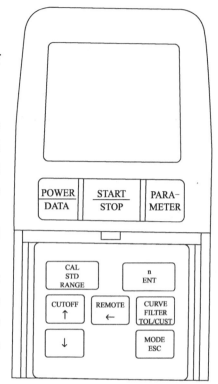

Fig.1-4-3　Operation panel

③ Measurement:

a. Place the SJ-201P surface roughness tester on the workpiece and confirm that the stylus is in proper contact with the measured surface.

b. Press the [START/STOP] key in the measurement mode to carry out the measurement. While measurement is being performed, "----" is displayed on the LCD. After the measurement has been completed, the measured value will be displayed on the LCD.

④ Switching the parameter to be displayed. Press the [PARA-METER] key until the desired parameter value is displayed. If the parameter displays setting of "Ra" currently, a displayed parameter is switched to "Ry", "Rz" and " Rq" in this order each time the key is pressed.

New Words and Phrases

nominal ['nɒmɪnl]　　　　　　　　　adj. 名义上的，票面上的，名字的

mating ['meɪtɪŋ]　　　　　　　　　　n. 配合

clearance [ˈklɪərəns]	n.	空隙，间隙
allowance [əˈlaʊəns]	n.	默许，默认
deviation [ˌdi:viˈeɪʃn]	n.	背离，偏离，偏差
straightness [streɪtnəs]	n.	正直，笔直，率直
flatness [flætnəs]	n.	平面
perpendicularity [ˈpɜ:pənˌdɪkjʊˈlerətɪ]	n.	垂直度，垂直，直立
vernier [ˈvɜ:nɪə]	n.	游尺，游标，游标尺
caliper [ˈkælɪpə]	n.	测径器，卡钳，弯脚器
micrometer [maɪˈkrɒmɪtə(r)]	n.	测微计，千分尺，目镜
rigid [ˈrɪdʒɪd]	adj.	僵硬的，严格的
sleeve [sli:v]	n.	套筒，套管
thimble [ˈθɪmbl]	n.	套管，套筒
anvil [ˈænvɪl]	n.	（铁）砧，测砧
ratchet [ˈrætʃɪt]	n.	（防倒转的）棘齿　　vi. 安装棘轮于，松脱
calibration [ˌkælɪˈbreɪʃn]	n.	校准，标准化，刻度，标度
whirligig [ˈwɜ:lɪgɪg]	n.	旋转，循环
directrix [dɪˈrektrɪks]	n.	准线
dimensional tolerance		尺寸公差
geometric tolerance		形位公差
profile of surface		面轮廓度
circular run out		圆跳动
vernier caliper		游标卡尺
outside micrometer		外径千分尺
surface roughness		表面粗糙度

Exercises

I. Match column A with column B.

A	B
校正值	nominal size
外径千分尺	allowance
允差	geometric tolerance
公称尺寸	cylindricity
形位公差	outside micrometer
圆柱度	surface roughness
表面粗糙度	calibration value

II. **Mark the following statements with T (true) or F (false).**

(   ) 1. Tolerance on a dimension can not specify the degree of accuracy.

(   ) 2. The dimension of the shaft is more than that of the hole, then it is termed interference fit.

(   ) 3. The vernier caliper provides the three basic functions of inner, outer and depth gauge.

(   ) 4. The vernier caliper is an even more rigid measuring instrument than outside micrometer.

(   ) 5. The process of calibration involves the measurement of a standard roughness specimen and the adjustment of the difference between the measured value and the standard roughness specimen.

III. **Answer the following questions briefly according to the text.**

1. What's definition of the clearance fit?
2. Which two kinds are tolerances divided into?
3. What is the application of the vernier caliper?
4. What does outside micrometer usually consist of?
5. What is the meaning of the symbol ?

Part B

Reading Material

Matters Needing Attention of Micrometer

Whenever you use a micrometer, carefully observe the DOs and DO NOTs in the following list to obtain accurate measurements and to protect the instrument:

① DO NOT measure moving parts because the micrometer may get caught in the rotating work and be severely damaged.

② Always open a micrometer by holding the frame with one hand and turning the thimble sleeve with the other hand.

③ Apply only moderate force to the thimble when you take a measurement.

④ When a micrometer is not in use, place it where it will not drop.

⑤ Before you store a micrometer, back the spindle away from the anvil, wipe all exterior surfaces with a clean, soft cloth, and coat the surfaces with light oil.

课文翻译

模块 1 机械技术基础

任务 1 机械工程

Part A

1.1.1 图　纸

1. 工程图纸

在机械制造中比较典型的机械图样有两种：零件图和装配图。

（1）零件图（图 1-1-1）

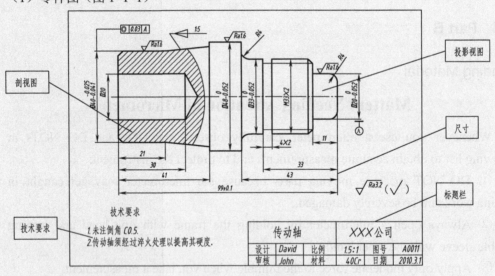

图 1-1-1 零件图

零件图是直接指导制造和检验零件的图样，一张完整的零件图包括一组图形、完整的尺寸、必要的技术要求和内容完整的标题栏。

① 一组图形：这组图形用一些必要的表达方法完整表达零件的内、外形状。

② 完整的尺寸：能够满足零件制造和检验时所需的尺寸。

③ 必要的技术要求：利用带有符号的标注或文字说明，表达出制造、检验和装配过程中应达到的一些技术上的要求，如表面粗糙度、尺寸公差，热处理和表面处理要求等。

④ 内容完整的标题栏：标题栏应包括零件的名称、材料、图号、图样的比例及图样的责任者签字等内容。

（2）装配图（图 1-1-2，略）

装配图是表达机器或组成部件的图样，在机械设计中，设计者首先画出装配图，具体表达所设计机器或部件的工作原理和结构，然后根据装配图分别绘制零件图。在机械制造过程中，首先根据零件图加工零件，然后按照装配图装配成部件或组装成机器。

2．机件的表达方法

国家《机械制图》标准规定采用视图、剖视图、断面图等各种方法来表达机件的结构和形状。

视图是机件向投影面投影所得的图形。最为典型的视图有基本视图、向视图、局部视图和斜视图。基本视图包括仰视图、主视图、俯视图、右视图、左视图和后视图，如图 1-1-3 所示。视图数目通常为六个、四个和三个，其中四个视图的情形最为典型。

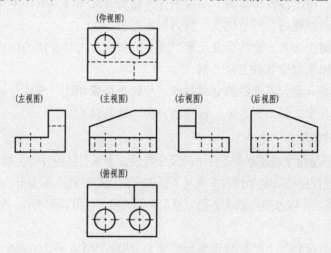

图 1-1-3　基本视图

剖视图（图 1-1-4，略）用来表达物体内部结构形状，且都要用虚线来绘制机件。当零件有复杂内部结构，而且清楚表达内部结构和外部结构同样重要时，可以采用"断面"技术切开零件的某断面来表达其内部结构。通常，把剖视图分为全剖视图、半剖视图和局部剖视图。

断面图（图 1-1-5，略）就是假想用剖切平面将机件的某处切断，仅画出断面的图形。断面图常见的有两种：移出断面图和重合断面图。

为了能够清楚地表达出零件的结构形状及大小，我们将零件放在一个三面投影体系中，利用正投影法将零件向三个投影面进行投影，得到三视图。三视图的基本投影规律是：主视图与俯视图长对正；主视图和左视图高平齐；俯视图与左视图宽相等。

为了说明形体的轮廓，还可以用轴测图。常见的轴测图有两种：正等轴测图和斜二等轴测图。

1.1.2 金属材料及其性能

1. 金属材料

金属材料分为两种基本类型：铁类金属和非铁金属。铁类金属主要有铸铁、碳钢、合金钢和工具钢。

铸铁主要分为三种：灰铸铁、白口铸铁和可锻铸铁。灰铸铁主要用于制造支架、汽车发动机气缸、手轮和机架；白口铸铁硬度高、耐磨性好，可以用来制造如火车车轮之类的零件；可锻铸铁是韧性较好，可用来制造如管道和扳手之类的零件，一般来说，铸铁有较好的抗压强度，较好的耐腐蚀性能和良好的机械加工性能，其主要缺点是脆性较大。

工业上使用的碳钢主要有三种：低碳钢、中碳钢和高碳钢。碳含量是影响碳钢力学性能的最重要因素。低碳钢含碳量为 0.05%~0.30%，主要用于制造强度要求不高的零件，如链条、螺钉、螺母和管道；中碳钢含碳量为 0.30%~0.50%，主要用于制造强度要求高于低碳钢的零件，如齿轮、曲轴、机器零件和车轴；高碳钢含碳量为 0.50%~1.70%，主要用于制造要求强度和硬度高的零件，如锉刀、刀、钻头、剃刀和木工刀。

合金钢是在碳钢中加入一定的合金元素以改变或提高其力学性能的钢铁材料。

合金工具钢是用来制造各种工具的钢。

非铁金属是主要合金元素非铁的金属材料。与铁类金属相比，非铁金属种类繁多。常用非铁金属有铝及铝合金、铜及铜合金、镁及镁合金、钛及钛合金。

2. 金属材料的性能

金属材料的性能是指金属在不同条件下的反应特性。金属材料的性能主要分为两类：物理性能和力学性能。物理性能是决定于材料本身且不能改变的固有性能，如重量、质量、颜色、相对密度。材料的力学性能可以改变，以满足某一要求，如强度、硬度、耐磨性、韧性、塑性和脆性。

（1）强度

强度是材料在载荷作用下抵抗塑性变形的能力。四种类型或形式的强度分别为抗拉强度、抗剪强度、抗压强度和极限强度。

（2）硬度

硬度是材料抵抗压痕或划痕的能力。测量金属材料硬度的方法很多，其中工业上主要应用的测试方法有两种：布氏硬度测试法和洛氏硬度测试法。

（3）耐磨性

耐磨性是材料抵抗磨损的能力。大多数情况下，材料硬度越高，耐磨性越好。

（4）韧性

韧性是材料抵抗或吸收或冲击载荷而不发生破坏的能力。

（5）塑性

塑性是材料产生塑性变形而不被破坏的能力。

（6）脆性

脆性是指材料在载荷作用下不产生明显变形而直接发生破坏的特性。脆性是与塑性相对的性能。

1.1.3 金属材料热处理

热处理是通过对金属进行加热和冷却控制,以使其性能得到显著改变的工艺方法。热处理工艺曲线如图 1-1-6 所示。钢铁材料最常用的五种热处理方法是:退火、正火、淬火、回火和表面淬火。大多数非铁金属可以进行退火,部分非铁金属可以进行淬火。但是,非铁金属一般不进行正火、回火或表面淬火处理。

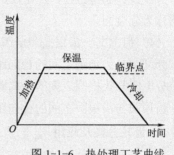

图 1-1-6 热处理工艺曲线

退火是用来软化金属的热处理工艺,一般对需要进行机械加工的硬化工件采用退火处理。退火的目的:降低硬度;提高可锻性;提高脆性;改善机械加工性能;改善组织。

正火是用来减少金属材料在加工或成形过程中产生的内应力的热处理工艺。正火的目的:消除应力;细化晶粒和改善结构均匀性;为机械加工做准备;减少热处理产生的变形。

淬火是用来提高金属材料强度、硬度和耐磨性的热处理工艺。淬火的目的:提高硬度、强度和耐磨性。

回火用来消除淬火时因为快速冷却而产生的应力。回火的目的:降低硬度到所需要的水平;提高抗冲击性和冲击强度;降低脆性;消除快速冷却产生的内应力。

表面淬火是通过对工件表面渗碳,使低碳钢工件表面获得硬化层的热处理工艺。表面淬火适合于要求表面硬度和耐磨性高,而心部韧性高的工件。典型的表面淬火工件有齿轮、链轮。

1.1.4 机械零件

1. 机器与零件

无论多么简单的机器,都是由单一构件即通称的机械零件或部件组成的。所以,假如把机器完全拆卸,就可以得到像螺母、螺栓、弹簧、齿轮、凸轮和轴等简单零件——这些是构成所有机器的标准组件。

机械零件中最常用的就是齿轮。齿轮将旋转运动从一个轴传递到另一个轴。通过改变主动齿轮和从动齿轮的尺寸,可以增加或降低旋转的速度。

2. 齿轮各部分的名称(图 1-1-7,略)

(1)齿数

在齿轮的圆周上均匀分布的轮齿总数,用 Z 表示。

(2)齿顶圆、齿根圆

相邻两齿间的空间称为齿槽,过所有齿槽底部的圆称为齿根圆,其半径用 r_f 表示。过所有轮齿顶部的圆称为齿顶圆,其半径用 r_a 表示。外齿轮的齿顶圆大于齿根圆,而内齿轮的齿顶圆小于齿根圆。

（3）分度圆

为使设计制造方便，人为规定一个圆，使该圆上的模数为标准值，其压力角也为标准值，该圆称为分度圆。

（4）齿顶高、齿根高和全齿高

分度圆与齿顶圆之间的径向距离称为齿顶高，用 h_a 表示。

分度圆与齿根圆之间的径向距离称为齿根高，用 h_f 表示。

齿顶圆与齿根圆之间的径向距离称为全齿高，用 h 表示。

3. 齿轮分类及应用

在现代机械中使用的齿轮有很多种，其中有直齿圆柱齿轮、斜齿轮、齿条、锥齿轮、蜗轮蜗杆等。

直齿圆柱齿轮（图 1-1-8，略）是齿轮中应用最广泛的一种类型，它用于在平行轴间传递旋转运动，并且保持恒定的速度和转矩。这种轮齿的渐开线齿廓是最易于产生的，并可获得很高的制造精度。

斜齿轮（图 1-1-9，略）与直齿圆柱齿轮相似，只是斜齿轮的轮齿是与轴的中心线成一定角度（螺旋角），这样可以产生更大的接触面积，从而可以承受更高的载荷和扭矩。

锥齿轮（图 1-1-10，略）仅用于两相交轴间传递旋转运动。

 Part B

退 火 类 型

退火是将金属稍微加热到临界温度以上后，很缓慢的冷却。退火处理能够减轻金属内部由于先前热处理、切削加工或其他冷加工所造成的内应力和应变。钢的种类决定着退火加热的温度，加热的温度也与退火的目的有关。

工业中应用的退火主要有三种：完全退火、低温退火和球化退火。

完全退火用来最大限度地降低钢的硬度，以改善它的切削加工性能，消除内应力。低温退火又称去应力退火，它的目的主要就是消除在冷加工和机械加工过程中产生的内应力。球化退火是使钢中生成一种特殊的晶粒结构，这种结构相对较软而易于加工。这种工艺一般用于改善高碳钢的切削加工性能。

任务2 机械制造

 Part A

1.2.1 金属切削加工技术

金属切削加工的六种基本方法包括：车削、铣削、刨削、磨削、钻削和镗削。

车削是金属加工的一种方法，是用切削刀具将金属表面多余的金属去除，生产出所需的产品，广泛用于车床中。在车削加工中，主轴的旋转运动是主运动，车刀的运动是进给运动。卧式车床如图 1-2-1 所示（图略）。

铣削是仅次于车削的应用最广的一种方法。立式、卧式铣床如图 1-2-2 所示（图略）。铣削时工件与有多条切削刃的旋转刀具相接触。针对不同的工件有多种铣床可供选择。铣床生产的某些形状相当简单，像由圆锯生产的缝槽和平面。其他更复杂的形状如平面与曲面的各种各样的组合，则是由给定刀具的切削刃的形状和刀具的运行轨迹得到的。

用机床刨削金属的过程类似于用手刨木头。其本质区别在于工件在刀具下面来回运动，而刨刀则固定在一个位置不动。龙门刨床通常用于加工大型工件。牛头刨床（图 1-2-3，略）与龙门刨床的区别在于它的工件是固定的而刀具是前后移动。

磨削加工是机械制造中重要的加工工艺。磨削刀具是旋转砂轮，砂轮就像有无数微切削的铣刀。这个操作常用于对经过热处理后变得很硬的工件进行精加工，得到精确尺寸和较好的表面质量。这是因为磨削能够纠正由热处理产生的变形。使用专用刀具在磨床上可以进行平面、外圆柱面和内圆柱面加工。近几年，磨削扩展了在金属强力切削方面的应用。万能磨床如图 1-2-4 所示（图略）。

钻削是在实体金属上加工通孔或盲孔的一种方法，刀具绕主轴旋转，相对工件进给。摇臂钻床如图 1-2-5 所示（图略）。钻削操作可以手动进行或在钻床上完成。一般在钻床上加工时，刀具绕主轴旋转而工件固定在工作台上不动。钻削是用旋转的钻头加工出一个圆孔。钻头可以有一个或几个切削刃及相应的排屑槽，排屑槽是直线形的或者是螺旋形的。排屑槽的作用是作为切削过程中排出切屑的通道及让润滑液、冷却液流向加工表面的通道。生产中使用较广的钻头有麻花钻、中心钻、枪钻和扁平钻。

镗削则是用一个旋转着的、偏置的单刃刀具对已经钻削出或铸造出的孔进行精加工。在某些镗床上，刀具固定而工件旋转；而在另一些镗床上，情况恰恰相反。镗床如图 1-2-6 所示（图略）。

1.2.2 刀 具

1. 刀具材料和性能

切削刀具的性能包括高硬度，特别是在加工过程中高温的情况下保持这种高硬度，还包括韧性、耐磨性，以及抗高压的能力。切削材料的选择应适合切削条件，如工件材料、切削速度、产品重量、选用的工作液等。常用的刀具材料有碳素钢、合金钢、高速钢、硬质合金和金刚石等。

2. 车削加工常用刀具

车削加工使用单刃车刀切削。常用的车刀有外圆车刀 [图 1-2-7 (a)]、切槽车刀 [图 1-2-7 (b)]、螺纹车刀 [图 1-2-7 (c)]、内孔车刀 [图 1-2-7 (d)]、麻花钻 [图 1-2-7 (e)] 和中心钻 [图 1-2-7 (f)] 等，（图 1-2-7 略）。

外圆车刀用于切削圆柱面、锥面和端面。切槽车刀用于分割工件和切断工件。螺纹车刀用于切削60°标准螺纹。内孔车刀用于加工内孔表面。麻花钻是孔的粗加工操作中最常用的钻头。中心钻为钻孔加工提供定位。

3. 铣削加工常用刀具

上表面铣削用面铣刀 [图 1-2-8（a）]；轮廓加工用键槽铣刀 [图 1-2-8（b）]；孔加工用中心钻 [图 1-2-8（c）]、麻花钻 [图 1-2-8（d）] 及绞刀 [图 1-2-8（e）]，（图 1-2-8 略）。

4. 刀具几何角度

刀具的切削部分包括切屑流出的表面和与工件接触的侧面。这两个相交的表面形成一个切削刃。刀具的性能取决于其所用材料和刀具角度，主要角度包括：刀尖角、前角、后角和偏角。

前角的大小决定了刀具的锋利程度，前角越大，刀具越锋利。前角可以是正的、负的或者为零。通常取值在0°到15°之间，而纵向前角通常取0°。

后角用来降低工件与后刀面之间的摩擦。后角的大小对工件表面质量有很大影响。同时，后角的大小决定了刀刃的强度，并能够影响切削刃的锋利程度。通常后角的取值在4°到6°之间（粗加工）或8°到12°之间（精加工）。

1.2.3 切削用量

1. 切削速度

铣削加工的切削速度为铣刀旋转时外边缘上一点的线速度。铣削加工时铣刀做旋转运动，铣刀的旋转速度称为主轴转速，用每分钟转数（r/min）来衡量。

每一把刀具都有自己的主轴转速，主轴转速的大小主要取决于被切削材料的种类和刀具的尺寸（直径）。切削相同的材料，刀具直径越小，主轴旋转速度越高。

车削加工的切削速度定义为工件边缘上刀具经过一点的线速度。它在数值上等于每分钟内工件边缘上的某一选定点沿切线方向移动的距离。切削速度与主轴转速之间的关系如下：

$$切削速度 = \pi D n / 1\,000$$

式中：D——工件外径，mm；

n——主轴转速，r/min。

每种材料都有一个合理的切削速度，这个速度是保证切削正常进行和获得较高表面质量的最佳切削速度。切削速度的大小主要取决于被加工材料和切削刀具的材质，这个数值可以从刀具制造商所提供的手册上查到。还有其他一些因素会影响切削速度的最佳值，例如，刀具的几何参数、润滑油和冷却液的种类、进给量和切削深度。

2. 进给量

刀具的进给量是指工件每转一周刀具与工件之间相对移动的距离。合适进给量大小的选择取决于许多因素，例如加工表面的光洁度要求、切削深度和刀具的几何参数。选取较小的进给量可获得较高的表面光洁度，选取较大的进给量能减少加工时间。因此，一般情况下常常选取较大的进给量进行粗加工，而选取较小的进给量进行精加工。从刀具制造商提供的手册中获得的进给量推荐值仅作为参考。

进给量通常用每分钟进给英寸数来测量。对于铣削加工，进给量指铣刀每分钟转过多少齿数。由于铣刀尺寸、齿数、每分钟转数的变化，因此进给量都以每齿进给量为基础来计算。在机床上设置进给量时必须把每齿进给量转换成每分钟进给量，进给量转换公式如下：

$$每分进给量（英寸/分钟）=主轴转速×每齿进给量×齿数$$

值得注意的是，每齿进给量太小会加速刀具的磨损，因此不是进给量越小越好。

3. 切削深度

切削深度是指刀具切入工件的深度，一般以 mm 为单位。对于车削加工，切削深度的计算公式如下：

$$a_p = (d_w - d_m)/2$$

式中：a_p——切削深度（图 1-2-9，略）；
　　　d_w——工件待加工表面直径（图 1-2-9，略）；
　　　d_m——工件已加工表面直径（图 1-2-9，略）。

确定切削深度必须考虑切削刀具、使用的机床和安装条件。切削深度直接影响刀具的寿命。如果切削深度太大，刀片将会过载而损坏。从手册中可以查到刀片的切削深度（a_p）的推荐值。如果切削深度太小，刀片受力会不均匀，会产生振动和不稳定。精加工时应选取较小的切削深度和选择刀具圆角半径小的刀具。

不正确地选择切削速度、进给量和切削深度将会影响工件的表面光洁度和引起机床或刀具损坏。选择切削用量时应考虑以下因素：

- 机床条件。
- 被加工材料类型。
- 工件在数控机床上的装夹方法。
- 切削刀具的类型。
- 刀具的直径。
- 切削刀具的材料类型。

Part B

切 削 液

1. 切削液的作用

① 减小磨损，以提高刀具的使用寿命和工件的表面粗糙度。
② 冷却加工区域，以提高刀具的使用寿命并降低工件的温度和热变形。
③ 减少切削力和能量的消耗。
④ 从加工区域带走切屑，以减少对加工过程的影响。
⑤ 保护加工表面。

2. 切削液种类

（1）合成油

合成油主要应用在低速加工的时候，因为温度对其影响不大。

（2）乳化液

乳化液是一种油、水和添加剂的混合溶液，通常使用在高速加工的场合，因为温度的升高对其影响较大。水的存在使得乳化液的冷却效果很好。

任务 3　数 控 编 程

Part A

1.3.1　数控编程基础

1. 坐标系（图 1-3-1）

机床制造商为每一台机床设置一个机床零点，机床零点是机床坐标系的原点。操作者编程时需要确定零件在机床上的位置时，必须建立工件坐标系。当工件尺寸太大时，用户可以在工件上的某一局部区域设定另一个坐标系，这个坐标系就是局部坐标系。

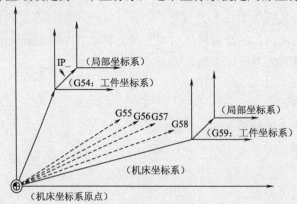

图 1-3-1　坐标系

坐标系中的坐标代表刀具的位置，坐标轴（图 1-3-2，略）用来确定坐标。3 个主要坐标轴称为 X、Y、Z 轴。为了建立右手直角坐标系 Z 轴垂直于 X 和 Y 轴，将刀具离开工件的方向定义为坐标轴正方向。

2. 程序零点

每个坐标轴的原点通常称为程序零点，又称工件零点、工件原点或零点。该点通常位于图纸上所有（或大部分）尺寸的起点。通常在程序的开始使用 G92（或 G50）命令设置程序零点。

程序零点的位置由编程者确定。程序零点可以设定在任何位置，合理选择程序零点会使编程更加简便。程序零点应选择在零件的尺寸基准上。每一个坐标轴都有一个程序零点。

车床程序零点一般设在工件右端面中点，如图 1-3-3 所示（图略）。

3. 绝对与增量坐标（图 1-3-4）

如果刀具移动到某一个目标点，该点的坐标值是以工件坐标系的原点为基准而确定的，此值即为该位置的绝对坐标值。如果刀具从当前点移动到某一个目标点，该目标点的坐标值是相对于刀具前一个位置坐标来计算的,此值即为该位置的增量坐标值。

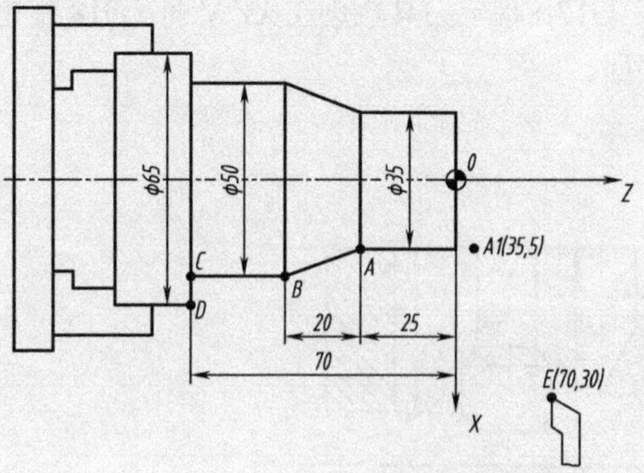

	绝对			增量	
点	X	Z	点	U	W
E	70	30	E	0	0
A1	35	5	A1	-35	-25
A	35	-25	A	0	-30
B	50	-45	B	15	-20
C	50	-70	C	0	-25
D	65	-70	D	15	0

图 1-3-4　运动的绝对与增量坐标

4. G、F、S、M 和 T 代码

G 代码又称准备代码或字。准备代码由地址 G 及其后面的 1 位或 2 位数字组成,它规定了数控机床沿其编程轴运动的模式。G 代码通常分为两类。模态 G 代码在随后的程序段一直有效,直至被同组的另一个代码代替;非模态 G 代码功能仅在所出现的程序段内起作用。

F 代码(图 1-3-5,略)控制切削进给速度。可以用每分钟进给量或每转进给量来表示。F 代码是模态的,在随后的刀具运动程序段中一直有效。一个程序中可以根据需要改变进给速度。

S 代码(图 1-3-6,略)控制主轴转速。S 地址后面是所选择的转速值。S 代码通常和主轴旋转指令(M03 或 M04)一起使用。

M 代码称为辅助功能字,控制机床的辅助功能。包括主轴转动/停止、机床的启动/停止、冷却液的开/关、换刀、返回程序头等功能。

T 代码(图 1-3-7,略)被用来指定刀具号。用于有自动换刀装置的机床。

1.3.2　手工编程代码

1. 基本编程代码

(1) G00(快速定位)

G00 指令是在工件坐标系中以快速移动速度移动刀具到达由绝对或增量指令指定的位置。在绝对指令中,用终点坐标值编程;在增量指令中,用刀具移动的距离编程。

(2) G01(直线插补)

G01 指令是将刀具以 F 代码指定的进给速度沿直线移动到指定的位置。F 代码中指定的进给速度一直有效,直到指定新值。不必对每个程序段都指定。

(3) G02/G03(圆弧插补)

该指令使刀具沿圆弧运动。圆弧圆心是用地址 I、J 和 K(分别对应于 X、Y 和 Z 轴的坐标)指定的。但是,I、J 或 K 后面跟的数值是从圆弧起点向圆心看的矢量分量。

（4）G17/G18/G19 （平面选择）

有些指令在被编程时必须要选择平面。例如，G17：指定刀具轨迹在 XY 平面；G18 指定刀具轨迹在 ZX 平面；G19 指定刀具轨迹在 YZ 平面。

2. 刀具补偿代码

（1）刀具长度补偿（图 1-3-8）

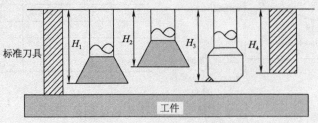

图 1-3-8　刀具长度补偿

在每一把刀具第一次 Z 向接近工件的运动中来设置刀具长度补偿。设置命令包括 G43 或 G44、指定相关刀具偏移量的地址 H，也必须有 Z 轴定位运动。一旦设置，刀具长度补偿将会一直有效，直到下一把刀具在第一次 Z 向运动时调用 G43 或 G44 指令（刀具长度补偿是模态的）。

G49 指令用来取消刀具长度补偿。

（2）刀具半径补偿（图 1-3-9）

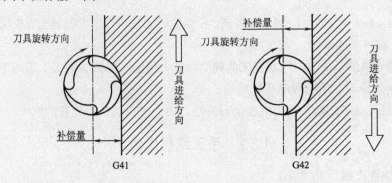

图 1-3-9　G41 与 G42

大多数的控制系统使用 3 种 G 代码形式的刀具半径补偿。G41 用来设置刀具左补偿（顺铣）。G42 用来设置刀具右补偿（逆铣）。G40 用来取消刀具半径补偿。此外，多种控制器在调用刀具半径补偿时，使用字符 D 来指定偏移量。

为了决定是调用 G41 还是 G42，可以简单地沿着刀具在加工过程中移动的方向去看，如果刀具在被加工表面的左侧调用 G41，如果在右侧调用 G42。一旦刀具半径补偿被正确设定后，刀具将会一直保持在所有需要加工表面的左侧或右侧，直到 G40 指令取消刀具半径补偿为止。

3. 循环代码

有多种类型的固定循环代码能简化编程。

（1）G71（粗车循环）

此循环指令能按照给定的精加工后零件形状粗车去除多余材料。通过编程精加工刀具轨迹来定义零件形状，并且通过 PQ 程序段指定把刀具轨迹赋予 G71 指令。

（2）G73（仿型粗车循环）

此循环指令可以车削固定的图形，并且是按此图形逐步逼近的。通过这种切削循环，就可以高效地切削已粗车成形、铸造成形或锻造成形的工件。

（3）G70（精车循环）

在 G71、G73 粗切后，G70 指令用来实现精加工。

（4）G92（简单螺纹切削循环）（图 1-3-10）

使用 G92 螺纹切削循环指令的好处是减少重复数据，使程序编制更加容易。在 G92 螺纹切削循环中，重要的是输入每一次螺纹切削后的直径。

（5）G81（钻孔循环）

当钻削浅孔时，编程员可以选择 G81 指令。

在使用 G81 指令编程时要输入一组信息：孔的 X 轴和 Y 轴坐标，Z 轴参考平面（R），和 Z 向最终加工深度。该指令能引起下列动作出现（见图 1-3-11）：

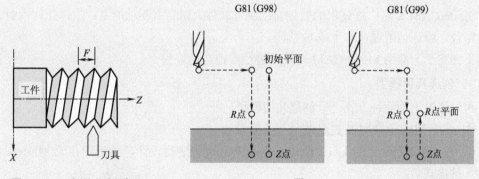

图 1-3-10　螺纹切削循环　　　　图 1-3-11　G81

① X 轴和 Y 轴快速定位到孔中心的位置上。

② 快速运行到 Z 向参考平面。

③ Z 向切削进给到最终深度。

④ 快速退回到 Z 向初始位置或 Z 向参考平面（R）。

（6）G80（取消钻孔固定循环）

钻孔固定循环取消后执行正常操作。G81 指令中被指定的 R 点和 Z 点被清除，其他的钻孔数据也被清除。

4. 子程序

如果一个程序包含固定顺序或频繁重复的图形，这样的顺序或图形就可以编成子程序存在存储器中以简化编程。子程序可以从主程序中调用，并且这个被调用的子程序还可以调用另外一个子程序。子程序调用如图 1-3-12 所示。

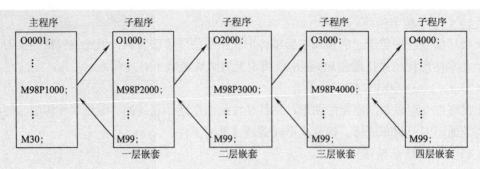

图 1-3-12 子程序调用

单个的子程序调用指令（M98）能重复调用一个子程序最高达 9 999 次。当主程序调用子程序时，它被认为是一级子程序调用，子程序调用最高可达四级嵌套。

子程序中 M99 指令被执行时，子程序就结束了，并且控制器将返回到调用程序段之后的一句或返回到由地址 P 指定的程序段号。

1.3.3 自 动 编 程

自动编程软件有很多，比如 MasterCAM、UG、Pro/E 等等。

MasterCAM X 是广泛使用的数控自动编程软件。它的主要功能是：创建几何模型、创建刀具路径、验证切削过程和生成 G 代码。

下面将学习应用 MasterCAM X 软件进行数控自动编程。

1．创建几何模型

可以通过下面两种方式之一创建几何模型：

① 使用 MasterCAM X 提供的交互绘图设计。

② 在 CAD 软件比如 UG、Pro/E、SolidWorks 中进行设计，然后保存成 MasterCAM X 能导入的格式。

2．创建刀具路径

（1）选择机床类型（图 1-3-13，略）

单击"机床类型"→"铣床"，选择机床类型。

（2）选择所需的刀具（图 1-3-14，图 1-3-15）

单击"刀具路径"→"刀具管理"，在"刀具管理"对话框中，显示所有可以选择的刀具。可以通过滚动条查找需要的刀具或者使用过滤按钮通过输入搜索信息进行查找。

（3）生成刀具路径（图 1-3-16）

① 单击"刀具路径"→"曲面粗加工"→"粗加工平行铣削加工"，弹出对话框，选择"零件形状"，单击 ✓ 按钮。

② 单击所要加工的零件表面，确保所有要加工的表面都被选择，然后单击 ✓ 按钮。

③ 设定粗加工平行铣削加工参数。

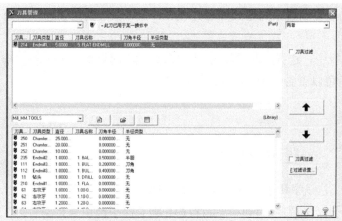
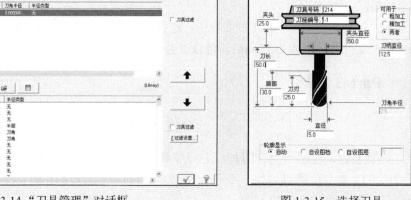

图 1-3-14 "刀具管理"对话框　　　　图 1-3-15 选择刀具

3. 验证切削过程［图 1-3-17、图 1-3-18（略）］

在一个零件加工之前，需要验证 CAM 模型的零件程序是否正确。验证的目的是为了：

① 检测刀具切削路径的几何错误。
② 检测刀具的潜在干涉。
③ 检测错误的切削条件。

验证开始前，在"刀具路径管理器"中选择一种或多种操作，然后单击验证按钮 ，使用"验证"对话框上部的控制按钮 进行仿真。

图 1-3-16 粗加工平行铣削加工参数设定　　　　图 1-3-17 "实体验证"对话框

4. 生成 G 代码（图 1-3-19，略）

不同的数控机床所使用的 G 代码形式有一定的区别。将加工的数据转换成某一机床所需的 G 代码格式称为后处理。G 代码被存储成不同格式的后处理文件，可以根据需要选择后处理文件格式。

从"刀具路径标签"的左侧面板中，选择 G1 图标，弹出"后处理"对话框，单击 ✓ 按钮，输入文件名。生成 G 代码后，可以查看 G 代码并可以进行修改。

📖 Part B

程 序 结 构

一般来说，程序类型有两种：主程序和子程序。主程序包括一系列加工工件的指令。子程序可以被主程序或另外一个子程序调用。不论是主程序还是子程序都由三部分组成：程序号、程序内容和程序结束。

程序按照程序号在 MCU 内存中存储。大多数数控机床可以同时存储多个不同的程序。程序号通常以"O"（或%）开始。"O"（或%）后面的数字是程序号。程序号的范围从 01 至 9 999。程序内容是整个程序的核心，由许多程序段组成。这些程序段控制数控机床要执行的运动。M02 或 M30 指令用于整个程序的结束。

一个数控程序包含一个或多个程序段。一个程序段是包含数控机床信息的一个完整行。它包含一个或一系列的字，程序段的长度可以变化。编程者必须掌握执行机床运动功能所需的字。

任务 4 公差及测量

📖 Part A

1.4.1 公 差

在图纸上标注尺寸时，尺寸线上标注的数字仅仅表示近似的尺寸，而不表示任何精度，除非设计人员加以说明。这个数字称为公称尺寸。零件的公称尺寸是设计人员根据设计工艺的需要而制定的一个恰当的尺寸值。然而，用已知的任一种制造工艺都几乎不可能将零件加工到百分之百精确的尺寸。因此，在工程实际中常允许在公称尺寸左右有一个变化范围，这称为公差。尺寸的公差也能说明尺寸的精确程度。例如，一根轴的公称尺寸为 33.5 mm，如果允许的变化范围是±0.05 mm，该尺寸就可以标注（33.5±0.05）mm。

在工程中当设计的某一产品是由许多零件组成时，这些零件以某种形式彼此配合。在装配时考虑到零件的配合类型是很重要的，因为这将影响到零件与零件之间的运动形式。以轴和孔配合为例，最简单的情况是轴的尺寸比孔的尺寸小，轴与孔之间会存在间隙，这样的配合成为间隙配合。反过来，若轴的尺寸比孔的尺寸大，则称为过盈配合。

一般来说，零件的成本随公差的减小而上升。如果一个零件要加工几个或更多的表面，允许的偏差与公称尺寸的差值很小时，那么成本将是非常大的。

允差有时与公差相混淆，两者的意义是完全不同的。允差是在两配合零件间允许的最小间隙，表示配合允许的最近的状态。假设直径为 $\phi 1.498_{-0.003}^{0}$ 的轴，要与直径为 $\phi 1.500_{0}^{+0.003}$ 的孔相配合，孔的最小尺寸为 1.500，轴的最大尺寸为 1.498，因此允差为 0.002，根据轴的最小尺寸和孔的最大尺寸可得最大间隙为 0.008。

公差分为两种：尺寸公差和形位公差。

尺寸公差可能是对称的，如尺寸 50±0.1；或者是非对称的，如 $\phi 50_{-0.045}^{-0.030}$，尺寸 50 是基本尺寸，上偏差为-0.030，下偏差为-0.045，因此，最大极限尺寸为 49.970。最小极限尺寸为 49.955。

形位公差用来指定形状和位置特征。比如：直线度、平面度、圆度、圆柱度、垂直度等。形位公差特征符号见表 1-4-1。

表 1-4-1 形位公差特征符号

种类	符号	种类	符号
直线度	—	倾斜度	∠
平面度	▱	平行度	∥
圆度	○	位置度	⊕
圆柱度	⌭	同轴度	◎
线轮廓度	⌒	对称度	≡
面轮廓度	⌓	圆跳动	↗
垂直度	⊥	全跳动	⌰

1.4.2 量具的使用

1. 游标卡尺的使用

游标卡尺（图 1-4-1）是一种测量仪器，由一个在其较长臂上刻有成比例的刻度的 L 形框架（称为主尺）以及一个带有游标尺的可滑动 L 形附件构成，用来直接读出由两个较短臂（称为卡钳）的内边或外边之间的间隔显示的某物的尺度。

游标卡尺有三种功能，分别是测量外部、内部和深度尺寸。

使用游标卡尺时，首先将卡钳夹在被测量物体上，要牢固卡住，但不能太紧。读数时，首先记录游标卡尺零刻度线在主尺上的位置，这是主尺上完整刻度线的数字，也就是要记录的整数部分。剩下的小数部分从游标尺上读出。这个数字是看游标上第几条刻度线与主尺的刻度线对齐。

无论使用什么型号的游标卡尺,都要采取以下防护措施,以免损伤游标卡尺。

① 在拿游标卡尺之前,洗净手上的油污,这些油污会损伤游标卡尺。

② 用前和用后擦净游标卡尺的各个组件。

③ 不要将游标卡尺掉落,那样会损坏游标卡尺。

2. 外径千分尺的使用

外径千分尺常简称千分尺,它是比游标卡尺更精密的长度测量仪器。图 1-4-2 是一个常见的量程为 0~25 mm 的外径千分尺。通常它由固定套筒、微分筒、测砧、测微螺杆和棘轮旋钮等组成。

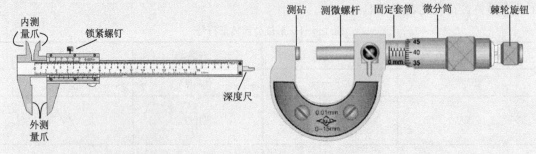

图 1-4-1 游标卡尺　　　　图 1-4-2 外径千分尺

固定套筒上有一条水平线,这条线上、下各有一列间距为 1 mm 的刻度线。上面的刻度线恰好在下面两相邻刻度线中间。微分筒被分成 50 等份,它是旋转运动的。当微分筒旋转一周时,测微螺杆前进或后退一个螺距——0.5 mm。

读千分尺时,把被测量的材料放在测砧和测微螺杆之间,然后转动棘轮旋钮直到测微螺杆和被测材料很好地接触。首先以微分筒的端面为准线,读出固定套筒下刻度线的分度值(只读出整数部分);再以固定套筒上的水平横线作为读数准线,读出可动刻度上的分度值。这两部分的和就是结果,前者为整数部分,后者为小数部分。

1.4.3 表面粗糙度检测仪

1. 表面粗糙度检测仪概述

SJ-201P 是一种适合车间现场使用的表面粗糙度检测仪,能够测量各种加工零件的表面。通过与粗糙度标准片对比,计算表面粗糙度值并显示测量结果。

2. SJ-201P 操作面板(图 1-4-3)

3. SJ-201P 测量操作

使用 SJ-201P 表面粗糙度检测仪测量的步骤

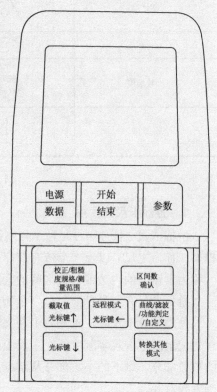

图 1-4-3 操作面板

如下：

① 按下 [POWER/DATA] 键，打开电源。

② 校正。校正是指通过测量粗糙度标准片，对测量值和标准值的差异进行调整的过程。下面是校正的过程：

a. 在测量模式下按 [CAL/STD/RANGE] 键，显示校正基准值。

b. 如果显示的数值与粗糙度标准片上标注的数值不同时，则需要修改校正基准值。如果数值不需要修改，则按 [ENT] 键，完成校正基准值的设定。

c. 安放粗糙度标准片。

d. 按 [START/STOP] 键开始校正的测量，在校正测量的过程中液晶屏幕上显示"-----"。当校正测量结束后，会显示测量值，按 [ENT] 键，完成校正操作。

③ 测量：

a. 将 SJ-201P 表面粗糙度检测仪放在工件上，确保探针正确地接触在测量面上。

b. 在测量模式下，按 [START/STOP] 键开始测量。在测量过程中液晶屏幕上显示"----"。当测量结束后，测量的结果会显示在液晶屏幕上。

④ 转换显示的参数。按 [PARAMETER] 键，直到显示所希望的参数值。如果当前显示的是"Ra"值，则每按下一次 [PARAMETER] 键，将会分别显示"Ry"，"Rz"和"Rq"的测量值。

 Part B

千分尺使用注意事项

当使用千分尺测量时，为了获得准确的测量结果和保护仪器，一定要仔细阅读下面的内容：

① 不能测量旋转的工件，这样会造成千分尺严重损坏。

② 一只手拿框架打开千分尺，另一只手旋转微分筒。

③ 测量时，旋转微分筒时力量要适当。

④ 千分尺使用完毕后，应正确放置以免摔落。

⑤ 储存千分尺前要将测微螺杆离开固定测砧，用布擦净千分尺外表面，抹上黄油。

Module 2 Foundation of Control Technology

Task 1 Electrical Control

 Part A

Text

2.1.1 Basic Concepts of Electrical Control

Electrical control refers to the use of electrical logic relations and operations to complete the control of the object of the task, that is, in order to achieve a certain purpose, to apply the required operation. Usually control tasks according to the needs of the differences, need to connect all the controlled amount and control device in accordance with a certain way to form an organism, that is, the electrical control system.

The electrical control system includes electrical control components, electrical protection components, electric actuator, electric circuit, mechanical transmission device, etc. In the design process it not only meets the basic principles of the design of electrical control system, but also fully considers the characteristics of mechanical equipment. Ultimately, it formates the simple, reliable, and economic control scheme.

2.1.2 Basic Law of Electrical Control

1. Point motion control

Point motion control is usually operated by the operator directly to control the signal (start button), to achieve the operation of the motor. When the start button is pressed, the motor runs; when the start button is released, the motor stops running. Point motion control mainly used to achieve the manual adjustment of production equipment, positioning, maintenance, and so on. Point motion control is shown in Fig. 2-1-1.

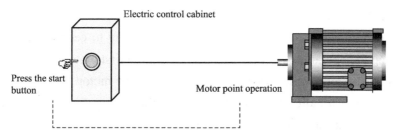

Fig. 2-1-1 Point motion control

2. Long dynamic control

In the long run, the operation of the motor is realized by controlling the starting signal. When the signal is switched on, the motor is running; when the signal is released from the start, because the circuit has the function of self lock so the motor continues to run, only press the stop button, the motor will stop running. Long dynamic control can realize the continuous operation of the motor, and the long dynamic control as shown in Fig. 2-1-2.

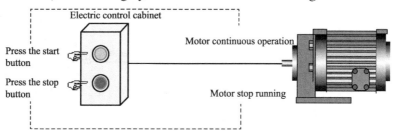

Fig. 2-1-2 Long dynamic control

3. Multi-point control

Multi-point control is that the operator can start the motor at different locations and stop operation. If the operator A through a press start or stop button in the cabinet, to achieve the motor running and stopping, the operator B can press the start or stop button in the production site, also can realize the operation of the motor run and stop. Multi-point control is generally used to control the production of a large number of field equipment or equipment stations. Multi-point control is shown in Fig. 2-1-3.

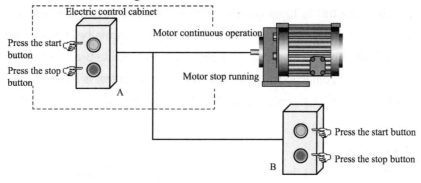

Fig. 2-1-3 Multi-point control

4. Sequence control

Sequence control refers to a number of motors according to certain order to implement more start or stop the operation. As shown in Fig. 2-1-4, the first motor can be started directly, the second sets of motors are allowed to start after the first motor started, the third sets of motors are allowed to start after the second sets of motors starting. In turn, third motors can be stopped directly, the second motors is allowed to stop after the third motors stopped, and the first motor is allowed to stop after the second motors stopped.

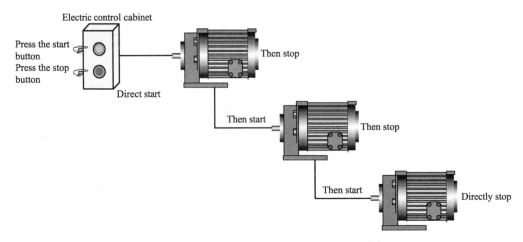

Fig. 2-1-4 Sequence of multiple motors' starting and stoping

5. Interlock control

The positive and reverse of the motor is required to implement the power supply commutation. In order to prevent the short circuit of the power supply, and the motor should be prohibited at the same time, positive and reverse power supply, this protective measure is interlocked. When the forward start button is pressed, the motor is running in the forward state, at this time, the reverse control should be shielded, vice versa. When you need to reverse the start, you should first make the motor to stop running, and then start the reverse run operation.

6. Button interlock control

Button interlock control is logic control relation that established between a plurality of control signals, such as the two buttons (normally open and normally closed) respectively to control the motor positively and reversely start and stop, you need to use two buttons (normally open and normally closed contacts) to cross. Button interlock control is also reflected in a set of signals to the other group of signals. Currently, the most use of control is button interlock control.

7. Stroke control

Stroke control is to control the travel and position changes of a moving part of production machinery. The control object is a stroke switch, which is sent out the signal by the mechanical force on the travel switch.

8. Time control

Time control is the use of time relay on the production of machinery and equipment to achieve timing control. As shown in Fig. 2-1-5, the feeding trolley stops running when it encounters the travel switch during traveling, after a certain period of processing (delay time) continue to the left, until the left limit switch to run after the stop. Time control is divided into the power delay and power off delay.

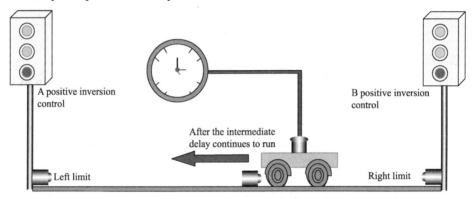

Fig. 2-1-5 Feeding trolley delay control operation

9. Condition control

Condition control refers to the electrical control system to meet the requirements of a certain condition after the implementation of the control. Control conditions are generally temperature, pressure, flow, liquid level, displacement and so on. Condition control is generally used for automatic production line, process control system.

2.1.3 Reading Electrical Diagrams

1. Classification of electrical diagrams

The electrical diagram is according to the standard drawing national electrical technical drawings, using electrical graphic symbols and text logo, and the provisions of the drawings and the form, performance of electrical equipment and system function, principle, and for the installation, maintenance of engineering drawings to provide technical data necessary. The electrical diagram mainly has electrical system diagram, electrical plan diagram, electric principle diagram, electrical components layout diagram, electrical wiring diagram.

(1) Electrical system diagram

Electrical system diagram is annotated with the symbol or frame, generally shows the basic components of system or subsystem, a sketch of mutual relations and main characteristics, as shown in Fig. 2-1-6.

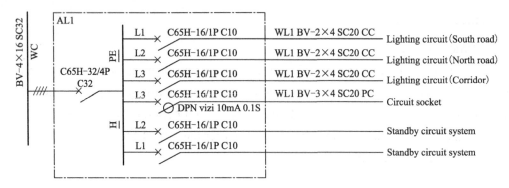

Fig.2-1-6　Lighting distribution system diagram

(2) Electrical plan diagram

The electrical plan diagram is a brief one which is based on the building plan to indicate the installation location of the equipment, installation and pipeline, and wiring and installation method, instead of reflecting specific shape, as shown in Fig. 2-1-7. Usually the pattern and the actual situation has certain scaling relations, which is to provide the installation of the main basis.

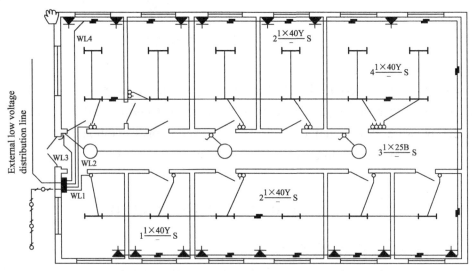

Fig. 2-1-7　Lighting distribution plan diagram

(3) Electric principle diagram

The electrical principle diagram is graphic symbols to represent the circuit, the working principle of equipment or system arranged according to the work order and reflect the logical relationship between them, regardless of its actual location of a diagram, as shown in Fig. 2-1-8. The electrical principle diagram is divided into main circuit and control circuit, which can be used to guide installation, connect wires, adjust, use and maintain.

(4) Electrical component layout diagram

Electrical component arrangement diagram is an electrical pattern that describes the layout and location of the device, as shown in Fig. 2-1-9. It includes in the cabinet and field distribution, such as the distribution of components in the electric control cabinet, distribution control device and discharge interval set, order device, installation and positioning. In the layout of electrical components, generally marked with the spacing of the components, the installation of the holes distance and the way out of the line.

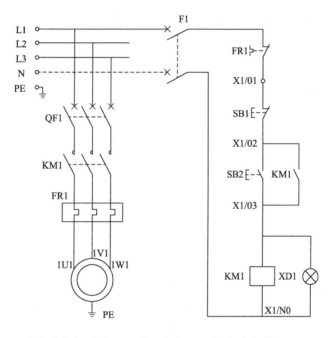

Fig. 2-1-8   AC motor electrical control principle diagram

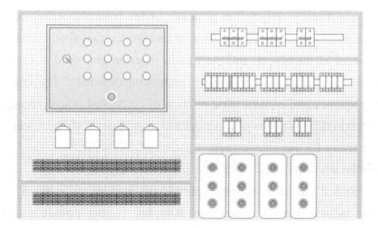

Figure 2-1-9   Layout of electrical components diagram

(5) Electrical wiring diagram

Electrical wiring diagram is a brief one to indicate the connecting relationship between the electrical equipment and components, which can be used to connect wires and check up, as shown in Fig 2-1-10.

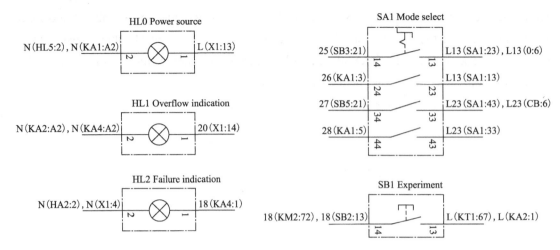

Figure 2-1-10   Electrical wiring diagram

2. Technique of electrical diagram analysis

The electrical control circuit mainly includes the main circuit, control circuit, auxiliary circuit and lighting circuit and so on. Before reading and analysis, you should firstly collection, organize, understand and analyze the overall structure of the object movement form, control process, control requirements, motor drive mode, transmission mode (mechanical, electrical, hydraulic), methods of operation, the installation position of motor and electrical components, working condition and requirement of automatic control and other technical information content.

(1) The main after the auxiliary of circuit analysis

The main circuit directly reflects the principle of mechanical transmission structure and action. If you read the main circuit of electrical principle diagram firstly, you can understand the electric equipment of the object, and their main function that need to controlled by what kind of appliances, and adopted what kind of protection measures. The control circuit can refine the action sequence and control logic relation of controlled object, through the analysis of the control circuit to determine the object's control requirements starting, steering, speed and braking, finally analyzes the auxiliary circuit.

(2) Break up the whole into parts of circuit structure analysis

The electric control circuit sometimes controls the content quite much, and the control logic is also quite tedious, but no matter how complex, control circuit is composed of the typical control link. The classification of functions and logic control of the relationship between the electrical control circuit, we often say that break up the whole into parts. Analysis of the circuit, we should start from the supply side, the master control switch to the contactor, relay coils, analysis one by one from top to bottom, from left to right, and pay attention to each local interlocking between the control circuit and interlocking relationship, combing the control of the process, express the working principle and process control circuit simply and clearly.

(3) Zero set of integrated analysis

After the initial break up the whole into parts process, you can understand the electrical control of each link, but in the end should be zero for the whole set of comprehensive analysis. To consider from the perspective of the whole control circuit, you should know the relationship between each control link, interlocking, interlocking relationship, clear mechanical, electrical, hydraulic setting coordination between various links and protection. To have a general understanding of the electrical control circuit, a further understanding of the electrical principle and process you should have, to understand and grasp each electrical circuits.

New Words and Phrases

organism ['ɔːgənɪzəm]	n. 有机体，生物体
transmission [træns'mɪʃn]	n. 传送，播送
positioning [pə'zɪʃnɪŋ]	n. 定位，配置，布置
maintenance ['meɪntənəns]	n. 维修，保养，保管
ultimately ['ʌltɪmətli]	adv. 最后，最终
cabinet ['kæbɪnət]	n. 电控柜
implement ['ɪmplɪment]	vt. 实施，执行
commutation [ˌkɒmjuˈteɪʃn]	n. 交换；折算
plurality [plʊəˈræləti]	n. 诸多，多元化
relay ['riːleɪ]	n. 继电器
trolley ['trɒli]	n. 手推车
implementation [ˌɪmplɪmenˈteɪʃn]	n. 成就，贯彻，满足
standard ['stændəd]	n. 标准，规格；adj. 标准的，合格的
legend ['ledʒənd]	n. 铭文；图例
layout ['leɪaʊt]	n. 布局，安排，设计；布置图，规划图
annotate ['ænəteɪt]	vt.& vi. 注解，注释
frame [freɪm]	n. 框架，边框
pipeline ['paɪplaɪn]	n. 管道；输油管道
auxiliary [ɔːgˈzɪliəri]	adj. 辅助的；备用的，补充的
electrical control	电气控制
manual adjustment	手动调整
point motion control	点动控制
long dynamic control	动控制
self lock	自锁
multi- point control	多点控制
sequence control	顺序控制

interlock control　　　　　　　互锁控制
button interlock control　　　　按钮联锁控制
stroke control　　　　　　　　行程控制
time control　　　　　　　　　时间控制

Exercises

Ⅰ. Match column A with column B.

A	B
长动控制	electrical control
自锁	layout
行程控制	maintenance
电气控制	long dynamic control
维修	self lock
布局	stroke control

Ⅱ. Mark the following statements with T (true) or F (false).

(　　) 1. In point motion control, the motor runs when the start button is pressed, and the motor stops when the start button is released.

(　　) 2. Multi-point control refers to the number of motors in a certain order to start or stop operation.

(　　) 3. In order to prevent the short circuit of the power supply, the motor should be allowed to be in the positive and reverse power supply at the same time.

(　　) 4. Electrical system diagram is annotated with the symbol or frame, generally shows the basic components of system or subsystem, a mutual relationship and the main features of the sketch.

(　　) 5. Electrical principle diagram is also called as the main circuit, debugging to guide the installation, wiring, use and maintenance.

(　　) 6. The layout of the electrical components is generally marked with the dimensions of each element, the mounting hole spacing, and the way of getting in and out.

Ⅲ. Answer the following questions briefly according to the text.

1. What is the electrical control?
2. What kinds of time control can be divided into?
3. What is the sequence control?
4. What are the conditions of control, the control requirement?
5. What are the skills in electrical diagram analysis?

Part B

Reading Material

Electrical Control System Design

Electrical control system design should have a certain design principles, and should ensure the controlled object's functionality, safety, rationality and economy, to combine the control process for the overall planning. The electrical control system should meet the requirements of mechanical equipment design of electrical control system; the electromechanical structure tends to be reasonable, complementary functions; control system is simple, economic, practical and reasonable selection of electrical components; reasonable selection of electrical components and layout plan; try to ensure the operation and maintenance convenience.

The design of electrical control system is mainly embodied in two aspects: control scheme design and control system design. The control scheme design is very important. It is the cornerstone of the design of the control system, only design a good scheme, design and implementation of effective talent. Electrical control system design should follow the principle of simple, efficient and practical. After the completion of the electrical control system design for its optimization, ultimately meet the design requirements.

Task 2 Hydraulic Control

 Part A

Text

2.2.1 The Work Principles of Hydraulics Transmission

Hydraulic transmission is a transmission form of transfer and control energy to use the liquid as the transmission medium, which can realize the transmission and automatic control of all kinds of machinery. Compared with mechanical transmission, hydraulic transmission is a new technology, which can use a variety of hydraulic components to build hydraulic circuit with different functions, and realizes energy transfer, conversion and control.

Fig.2-2-1 shows the real product and working principle of the hydraulic jack. Its structure is mainly composed of the manual plunger pump (piston1, lever2), hydraulic cylinder, check valve, shutoff valve, reservoir, etc. According to Pascal's law, pressure exerted on a confined liquid is transmitted undiminished in all directions and acts with equal force on all equal areas. The analysis of working process is as follows:

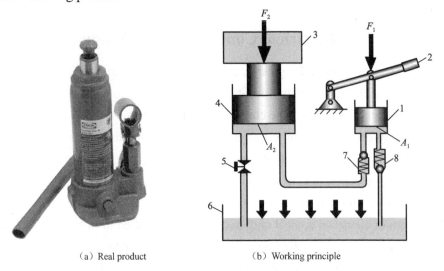

(a) Real product   (b) Working principle

Fig.2-2-1 Real product and working principle of the Hydraulic jack

1-piston; 2- lever; 3-load; 4-hydraulic cylinder; 5- shutoff valve; 6-reservoir; 7,8-check valve

1. Power transmission

The hydraulic pressure in the hydraulic cylinder P_2: $P_2 = \dfrac{F_2}{A_2}$

The thrust on the small piston of pump F_1: $F_1 = P A_1 = P_1 A_1 = P_2 A_1 = \left(\dfrac{A_1}{A_2}\right) \cdot F_2$

We can get the first important law here that the working pressures in the hydraulic system depended on the outside load.

2. Motion transmission

The volume of displacement is equal to the volume drawn into it, and then:

$$V_1 = v_1 t \cdot A_1 = v_2 t \cdot A_2 = V_2$$

Dividing by the moving time t on the two sides in the above formula, we have:

$$q_1 = v_1 A_1 = v_2 A_2 = q_2$$

We can get the second important law here that the motion speed of piston in actuator depends on the inlet flow rate, and independent of the outside load.

From the above analysis, hydraulic oil pressure and the flow rate are the two main parameters in the hydraulic transmission.

3. The applied examples of hydraulic transmission

Fig. 2-2-2 shows the hydraulic system of the grinding machine workbench. The main components of hydraulic system are hydraulic pump, hydraulic cylinder, flow control valve, directional control valve(reversing valve), pressure control valve, relief valve, filter, reservoir, etc. While working, the overflow valve is used to adjust or limit the pressure of the hydraulic system, flow control valve can adjust the size of the hydraulic cylinder speed, three-position four-way directional control valve can make the workbench movement or stopped. Specific hydraulic circuits are as follows:

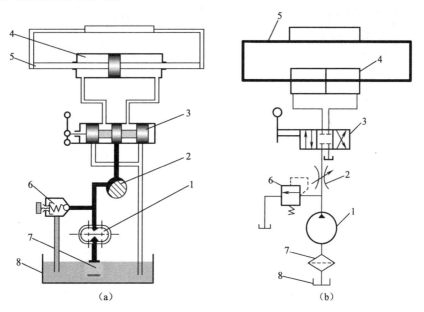

Fig.2-2-2 Hydraulic system of the grinding machine workbench
1-hydraulic pump; 2- flow control valve; 3- directional control valve; 4-hydraulic cylinder;
5- workbench; 6-relief valve; 7-filter; 8-reservoir

When the left-position of directional control valve is on work, the oil in the hydraulic from tank→filter→hydraulic pump→throttle valve→reversing valve (left-position) →the hydraulic cylinder (left-cavity), drives the workbench to the right direction.

When the right-position of directional control valve is on work, the oil in the hydraulic from tank→filter→hydraulic pump→throttle valve→reversing valve (right-position) →the hydraulic cylinder (right-cavity), drives the workbench to the left direction.

When the middle-position of directional control valve is on work, the oil in the hydraulic from tank→filter→hydraulic pump→throttle valve→reversing valve (middle-position), the workbench stops movement due to the O-function of directional control valve.

2.2.2 The Composition of Hydraulic Transmission System

Any hydraulic transmission system can be divided into five logical segments:

1. Energy portion

Energy portion can convert mechanical energy to the liquid pressure energy, and provides power for hydraulic system, such as hydraulic pump (Fig. 2-2-3).

Fig.2-2-3　Hydraulic pump

2. Execute element

Execute elements can convert liquid pressure energy to mechanical energy and drive different mechanical device. Such as hydraulic cylinder (Fig. 2-2-4) which can be used for linear motion and hydraulic motor (Fig. 2-2-5) which can be used for rotational motion.

Fig.2-2-4　Hydraulic cylinder　　　　　　　Fig.2-2-5　Hydraulic motor

3. Control element

It can control and regulate the pressure, flow and flow direction of liquid in hydraulic system. Such as overflow valve (Fig. 2-2-6), directional control valve (Fig. 2-2-7), flow control valve (Fig. 2-2-8), etc.

Fig.2-2-6　Overflow valve　　Fig.2-2-7　Directional control valve　　Fig.2-2-8　Flow control valve

4. Assistant element

Assistant elements can guarantee the normal work of the system outside the above three kinds of device. Such as reservoir (Fig.2-2-9), filter (Fig.2-2-10), energy accumulator (Fig. 2-2-11), etc.

Module 2　Foundation of Control Technology

Fig.2-2-9 Reservoir　　　　Fig.2-2-10 Filter　　　　Fig.2-2-11 Energy accumulator

5. Working medium

Working medium plays the role of energy transfer, lubrication, antiseptic, antirust, cooling, etc. Such as hydraulic oil.

In order to simplify the representation of hydraulic transmission systems, graphic symbols are usually used to draw the schematic diagram of the system. Fig.2-2-2 (b) is the hydraulic transmission system schematic diagram of Fig.2-2-2 (a).

2.2.3　Characteristics and Application of Hydraulic Transmission

1. The advantages of hydraulic transmission

① The hydraulic transmission system has smaller volume, light and compact configuration at a given power.

② The hydraulic transmission system has a good working stability.

③ The hydraulic transmission system can reach a wide range of speed regulation.

④ The hydraulic transmission can be easy realized automation.

⑤ The hydraulic transmission system is easy to realize overload protection.

⑥ The hydraulic transmission system is easier in design, fabrication and application.

⑦ The hydraulic transmission system is easier than machine equipment in doing line motion.

2. The shortages of hydraulic transmission

① Oil leaks are inevitable.

② Working performance is susceptible to the influence of temperature change.

③ Hydraulic components manufacturing precision demand is higher, so the price is more expensive.

④ It is difficult to find the reasons of fault.

3. The application of hydraulic transmission

The application of hydraulic transmission technology in industries as shown in Table 2-2-1.

Table 2-2-1   The application of hydraulic transmission technology in industries

Fields name	Examples
Engineering machine	Grab, loading machine, bulldozer, shovel machine, etc.
Mine machine	Charge, digger, elevator, hydraulic support, etc.
Architecture machine	Pile driver, hydraulic jack, flat machine, etc.
Metallurgy machine	Rolling mill, press machine, etc.
Manufacturing	Tool machine, CNC machining center, automatic assembly line, press machine, model-forge machine, etc.
Light industriy	Packer, injection-plastic machine, food packager, etc.
Automobile industry	High altitude operating car, truck crane, redirector, etc.
Water project	Dam, strobe, ship machine, ship-rudder, etc.
Farming industry	Fertilizer packager, combine harvester, tractor, farming suspension system, etc.

New Words and Phrases

hydraulic [haɪˈdrɔːlɪk]	adj. 水力的，水压的
transmission [trænsˈmɪʃn]	n. 播送，传送，传动装置
workbench [ˈwɜːkbentʃ]	n. 工作台，作业台
cylinder [ˈsɪlɪndə(r)]	n. 圆柱，圆筒，圆柱体，气缸
pump [pʌmp]	n. 抽水机，打气筒，泵 v. 用泵输送，涌出
valve [vælv]	n. 真空管，阀门，电子管
reservoir [ˈrezəvwɑː(r)]	n. 蓄水池，贮液器，储藏
filter [ˈfɪltə(r)]	n. 滤波器，滤光器，[化] 过滤器
	vi. 过滤，透过，渗透 vt. 过滤，滤除
parameter [pəˈræmɪtə(r)]	n. 因素，特征，参量
hydraulic transmission	液压传动
Pascal's law	帕斯卡定律
hydraulic jacks	液压千斤顶
grinding machine	磨床
mechanical energy	机械能
liquid pressure energy	液体压力能
right-cavity	右腔
directional control valve	换向阀
actuator element	执行元件
overload protection	过载保护
check valve	单向阀
shutoff valve	截止阀
relief valve	安全阀
overflow valve	溢流阀

Exercises

Ⅰ. Match column A with column B.

A	B
机械能	hydraulic pump
液压泵	flow control valve
溢流阀	directional control valve
方向控制阀	hydraulic cylinder
流量控制阀	overflow valve
液压缸	mechanical energy

Ⅱ. Mark the following statements with T (true) or F (false).

( ) 1. Hydraulic pump is actuator element.
( ) 2. Relief valve can control the liquid pressure of hydraulic system.
( ) 3. Hydraulic system can't reach a wide range of speed regulation.
( ) 4. Hydraulic components manufacturing precision demand is higher.
( ) 5. Hydraulic cylinders can be used for rotational motion.

Ⅲ. Answer the following questions briefly according to the text.

1. What is the main parameters of the hydraulic system?
2. What is the function of the hydraulic pump?
3. What is the working principle of the hydraulic system?
4. What are the components of the hydraulic system?

 Part B

Reading Material

The Operation and Maintenance of Hydraulic System

① The operator must be familiar with operating essentials of hydraulic components, the relationship between the rotation direction of handle and the size of pressure or flow, and beware of hydraulic accidents.

② Pay attention to oil level and temperature at any time while working. Generally the normal working temperature of the oil is 30 ~ 60 ℃, and the highest temperature should not exceed 60 ℃. We should stop it for checking abnormal temperature when abnormal temperature rises. Heater should be used when temperature low in winter.

③ Hydraulic oil should be regularly inspected and replaced to maintain it clean. For new hydraulic equipment, the tank should be cleaned and the oil should be replaced when the hydraulic oil is used for three months, after every six months to a year for a cleaning and oil.

④ Pay attention to the use of oil filter, filter must be regularly cleaned and replaced.

⑤ If the hydraulic equipment is not used for a long time, all the regulating handle should be loosened to prevent the spring from producing permanent deformation.

Task 3　　Pneumatic Control

 Part A

Text

2.3.1　The Working Principles of Pneumatic Transmission

Pneumatic transmission and control technology referred to as pneumatics technique. Pneumatic and hydraulic transmissions are similar in operating principle and control means. It is a transmission form which uses compressed air as transmission medium to transfer and control energy or signal, and is also a technology to realize the automation of production process by controlling and driving of all kinds of machinery and equipment.

Fig.2-3-1 shows the working principle of pneumatic shearing machine, the flow path of compressed air in pneumatic system is as follows: air compressor 1→cooler 2→oil-water separator 3 (cooled and preliminary purification) → gas tank 4 (spare) → water-separating gas filter 5 (again cleaning) → pressure reducing valve 6 (regulating and stabilizing pressure) → air lubricator 7 (lubricated components) → reverse valve 9 →air cylinder 10.

The specific working process of pneumatic shearing machine is as shown in Fig.2-3-2. Fig. 2-3-2(a) is the preparation state before shear. At this moment, the stroke valve is closed, the upper position of the reverse valve is in the working position, the piston of the cylinder moves downward, and the shear blade is opened.

Fig. 2-3-2(b) is the shear state. When workpiece is sent into the fixed position of the shearing machine, the stroke valve is pressured and connected with the atmosphere, the valve core of reversing valve moves down, the under position of the reverse valve is in the working position, and the shear blade which is driven by cylinder piston will rapid upward movement and cut the workpiece.

When the workpiece is cut off, the stroke valve reset, the valve core of reversing valve moves up and also reset. The cylinder piston drives the shear blade downward movement, and the system returns to preparation state as Fig. 2-3-2(a) for the second feed cut.

Module 2 Foundation of Control Technology

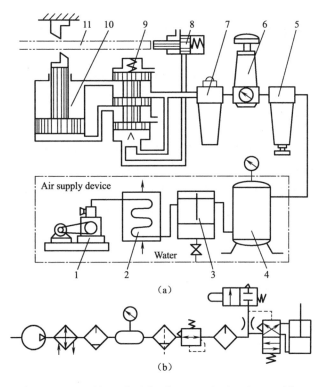

Fig.2-3-1 Working principle of pneumatic shearing machine

1-air compressor; 2-cooler; 3-oil-water separator; 4-gas tank; 5-water-separating gas filter; 6-pressure reducing valve;
7-atomized lubricator; 8-stroke valve; 9-reverse valve; 10-air cylinder; 11-workpiece

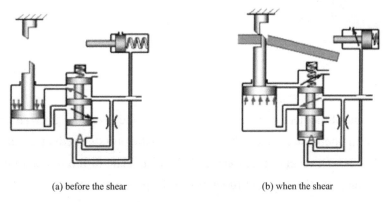

(a) before the shear   (b) when the shear

Fig. 2-3-2 working process of pneumatic shearing machine

From the above examples, we can draw conclusion: pneumatic transmission, which uses compressed air as working medium, can be used to transmit motion and power through gas pressure energy in closed volume. The essence of pneumatic transmission is energy transformation between mechanical energy and gas pressure energy. i.e., "mechanical energy→gas pressure energy→mechanical energy".

2.3.2 The Composition of Pneumatic Transmission System

According to the working principle diagram of the pneumatic shearing machine, a complete pneumatic transmission system has the following components:

1. Air supply device

As shown in Fig. 2-3-3, air supply devices are used to provide compressed air (with a certain lever of pressure, flow rate and purification) for pneumatic transmission systems. It includes the following parts: air compressor which can convert the mechanical energy to the gas pressure energy, device and equipment for purifying and storing compressed air, and air piping systems for transmitting compressed air, etc.

2. Pneumatic actuator elements

Pneumatic actuator element can convert the gas pressure energy of air compressor to the mechanical energy and drive different mechanical device. It includes air cylinders (Fig. 2-3-4) which can be used for linear motion and air motors (Fig. 2-3-5) which can be used for rotational motion.

Fig. 2-3-3   Air supply device   Fig. 2-3-4   Air cylinder   Fig. 2-3-5   Air motor

3. Pneumatic control elements

Pneumatic control element can control and regulate the pressure, flow and flow direction of compressed air in pneumatic system. Such as pneumatic pressure reducing valve(Fig. 2-3-6), throttle valve(Fig. 2-3-7), reversing valve(Fig. 2-3-8), etc. The different combinations of these components can complete the different functions of the pneumatic system.

Fig. 2-3-6   Pressure reducing valve   Fig. 2-3-7   Throttle valve   Fig. 2-3-8   Reversing valve

4. Pneumatic auxiliary elements

Pneumatic auxiliary element can connect pneumatic components, eliminate noise, cool, measure, etc. Such as pipe, pressure gauge (Fig. 2-3-9), filter (Fig. 2-3-10), muffler (Fig. 2-3-11), oil mist device (Fig. 2-3-12), etc. They play a very important role in maintaining the normal, reliable and stable operation of the system.

Fig. 2-3-9 Pressure gauge   Fig. 2-3-10 Filter   Fig. 2-3-11 Muffler   Fig. 2-3-12 Atomized lubricator

5. Working medium

Working medium plays the role of energy transfer in pneumatic system. Such as compressed air.

2.3.3 Characteristics and Application of Pneumatic Transmission

1. The advantages of pneumatic transmission

① The air can be obtained and expelled easily from the atmosphere.

② The air has low viscosity and lower pressure loss in pipes. It's very convenient to centralized supply and transmitted over a long distance.

③ Requirements for materials and manufacturing accuracy of pneumatic components are relatively low.

④ The maintenance of pneumatic system is simple, and the pipeline is not easy to block.

⑤ The use of pneumatic system is safe, and it is easy to achieve overload protection.

2. The shortages of pneumatic transmission

① The working stabilities are poorer than those of hydraulic transmission system.

② The push force of pneumatic transmission is usually very lower, so the transmission efficiency is lower.

3. The application of pneumatic technology

Due to the working medium of pneumatic transmission is compressed air, which has the characteristics of fireproofing, explosion-proof, anti-electromagnetic interference, anti-vibration, non-pollution, simple structure and reliable work, etc., so the combination of pneumatic technology and hydraulic, mechanical, electrical technology has developed into an important method to achieve the automation of production process.

Pneumatic technology is now widely used in machinery, electronics, light industry, textiles, food, medicine, packaging, metallurgy, petrochemical, aviation and transportation and other industrial sectors. Such as pneumatic manipulator, combination machine tools, machining centers, automatic production line, automatic detection, experimental device, etc. Pneumatic technology shows great superiority in improving production efficiency, automation, product quality, work reliability and other aspects.

New Words and Phrases

pneumatic [njuːˈmætɪk]	*adj.* 充气的，气动的，装满空气的
similar [ˈsɪmələ(r)]	*adj.* 类似的，同类的，相似的 *n.* 类似物，相似物
compress [kəmˈpres]	*vt.* 压紧，压缩 *n.* 打包机
separator [ˈsepəreɪtə(r)]	*n.* 分离器，分离装置
purification [ˌpjʊərɪfɪˈkeɪʃn]	*n.* 洗净，提纯
lubricate [ˌluːbrɪkeɪt]	*vt.* 加油润滑，使润滑 *vi.* 润滑
actuator [ˈæktʃʊeɪtə]	*n.* 激励者，执行机构
linear [ˈlɪnɪə(r)]	*adj.* 直线的，线形的，长度的
auxiliary [ɔːgˈzɪlɪəri]	*adj.* 辅助的，备用的，补充的
maintenance [ˈmeɪntənəns]	*n.* 维持，保持，保养，维护，维修
manipulator [məˈnɪpjuleɪtə(r)]	*n.* 操作者，操纵者，操纵器
air compressor	空气压缩机
oil-water separator	油水分离器
gas tank	储气罐
stroke valve	行程阀
pressure reducing valve	减压阀
shear blade	剪刃
gas pressure energy	气压能
explosion-proof	防爆
electromagnetic interference	电磁干扰
regulating and stabilizing pressure	调压和稳压
pressure gauge	压力表
oil mist device	油雾器

Exercises

Ⅰ. Match column A with column B.

A	B
气压能	pressure reducing valve
防火	stroke valve
气动技术	gas pressure energy
行程阀	air cylinder
减压阀	pneumatic technology
气缸	fireproofing

Ⅱ. Mark the following statements with T (true) or F (false).

(　) 1. Air motors are actuator element.

(　) 2. Pressure reducing valve can't control the air pressure of pneumatic system.

(　) 3. The pipeline is easy to block in pneumatic system.

(　) 4. The air has low viscosity and lower pressure loss in pipes.

(　) 5. The push force of pneumatic transmission is usually very higher.

Ⅲ. Answer the following questions briefly according to the text.

1. What is the function of the air compressor?
2. What is the working principle of the pneumatic system?
3. What is the advantages of pneumatic transmission?
4. What are the components of the pneumatic system?

 Part B

Reading Material

Fault Diagnosis Method for Pneumatic System

1. Experience method

Experience method refers to the fault diagnosis and elimination methods based on the practical experience and simple instrument, which can be realized through traditional Chinese medicine (TCM) diagnosis mode "look, smell, ask, feel".

2. Reasoning analysis method

The reasoning analysis method can find out the real reason of fault according to the logical reasoning.

(1) Instrument analysis method

Detect the technical parameters of the system or components with the instruments and meters, and determine whether it meets the requirements of pneumatic system.

(2) Partial stop method

Temporarily stop the work of a certain part of the pneumatic system, and observe the impact on the fault symptoms.

(3) Apagoge test

Tentative change part of the working conditions in the pneumatic system and observe the impact on the fault symptoms.

(4) Comparison method

Replace the same components in the system with the standard or qualified components, and determine whether the replaced elements are invalid through comparing working condition.

Task 4　PLC Control

 **Part A**

Text

2.4.1　The Definition of PLC

Programmable logic control (PLC) is a kind of electronic systems with digital computing operation, which is specially designed for application in industrial environment. It adopts the programmable memory, which is used to store instructions in its internal space to perform logical operations, sequence control, timing, counting, arithmetic operations, and controls various types of mechanical production process through digital and analog input and output.

PLC has the characteristics of process oriented, user oriented, strong adaptability to the industrial environment, convenient operation and high reliability. Its control technology represents the advanced level of process control, and has become basic automatic control system equipment.

2.4.2　The Composition of PLC

PLC is a kind of general industrial control computer, which is composed of two parts of hardware and software.

1．The hardware composition of PLC

There are many kinds of PLC type, but the structure and working principle are basically the same. Fig.2-4-1 shows the composition of PLC, it consists of five parts: central processing unit(CPU), memory, input/output unit, power unit, programming unit.

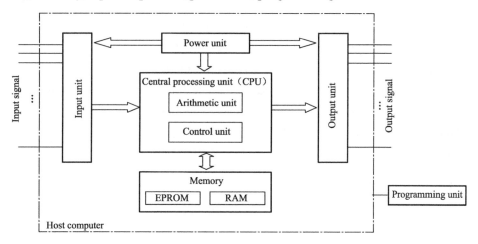

Fig. 2-4-1 Composition of PLC

(1) Central processing unit (CPU)

CPU is the core component of PLC, similar to the brain, which includes arithmetic unit and control unit.

(2) Memory

Memory is used to store data or programs, including random access memory (RAM), read-only memory (ROM), erasable programmable read only memory (EPROM).

(3) Input/output unit

Input/output unit is interconnected channels for PLC and I/O devices or other external devices and can input/output all kinds of operation level and drive signal.

(4) Power unit

Power unit is a switch type voltage stabilized power supply for the internal circuit of the PLC.

(5) Programming unit

Programmer unit is the main external equipment, which can realize the compiling, editing, debugging and monitoring of user program, and can also call and display the PLC's internal state and system parameters through the keyboard. Therefore, it is also a main tool for system operation and fault analysis.

2. The programming language of PLC

(1) Sequential function chart (SFC)

(2) Ladder diagram (LAD)

(3) Statement table (STL)

(4) Function block diagram (FBD)

(5) High-level language

2.4.3　The Working Process of PLC

As shown in Fig.2-4-2, PLC's work is continuous cycle scanning process. The whole scan process includes five stages: internal processing, communication processing, input scan, execute user program and output processing. The time required to scan the entire process is called the scan cycle, the length of the scan cycle is related to the length of the user program and the speed of the scan.

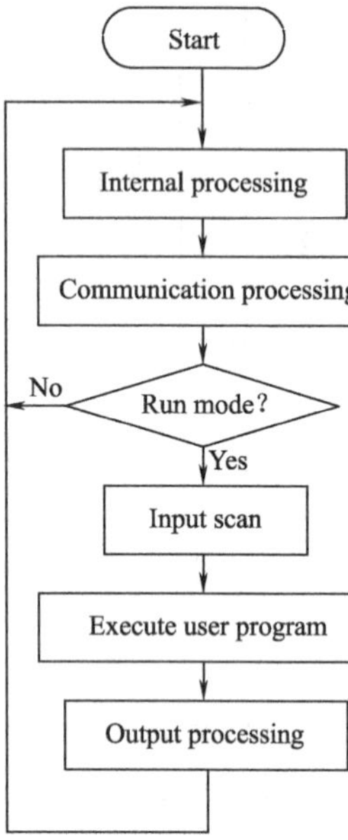

1. Internal processing

Internal processing includes system initialization and self diagnosis test.

2. Communication processing

It mainly completes the information exchange between PLC, PLC and the host computer or terminal equipment.

3. Input scan

Before executing the user program, the PLC reads all the state of the input terminal in a scanning mode, and is stored in the input image register. At this moment, the contents of the input image register are refreshed.

Fig.2-4-2　Working process of PLC

4. Execute user program

When executing the user program, CPU executes program instructions in sequence in scanning way from the first instruction, until the last instruction.

5. Output processing

After executing all program instructions, PLC will save the state of all the output relays in the component image register in a batch mode to the output latch register, and drive an external load through a certain output way.

2.4.4　The Applied Examples of PLC

Fig.2-4-3 shows working principle diagram of umbrella testing machine, which can continuous analog the movement of automatic open umbrella, and complete the times test of consecutive opening umbrella without breakdown.

It's working condition requirements as follows:

Module 2 Foundation of Control Technology

In the initial state, piston rod of cylinder 1 is retracted, piston rod of cylinder 2 is pushed out.

When working,press the start button→1YA "+" →piston rod of cylinder 2 is retracted(close umbrella) →magnetic switch B1 is pressured→2YA"+"→piston rod of cylinder 1 is pushed out(press the button of open umbrella)→delay 1 second→1YA"-" →piston rod of cylinder 2 is pushed out(open umbrella)→2YA"-" →piston rod of cylinder 1 is retracted(loosen the button of open umbrella),the above is an test of opening umbrella.

How to control the pneumatic system by PLC? The following is the design process of the PLC system.

1. Draw the electrical control diagram of umbrella testing machine

According to the working process of umbrella testing machine, we can draw the electrical control diagram as shown in Fig.2-4-4. Its control process is as follows:

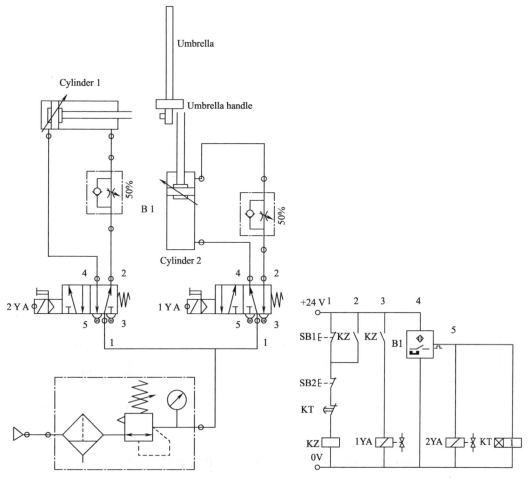

Fig. 2-4-3 Working principle diagram of umbrella testing machine

Fig.2-4-4 Electrical control diagram of umbrella testing machine

Press the start button SB1→electric relay KZ "+" →solenoid valve 1YA "+" (piston rod of cylinder 2 is retracted)→magnetic switch B1 is connected→solenoid valve 2YA "+" (piston rod of cylinder 1 is pushed out)→ time relay KT"+"(delay 1 second)→solenoid valve 1YA "-"(piston rod of cylinder 2 is pushed out)→magnetic switch B1 is disconnected→solenoid valve 2YA "-"(piston rod of cylinder 1 is retracted).

2. Assign I/O address

According to the requirement of the PLC system, the assignment of I/O address as shown in Table 2-4-1.

Table 2-4-1 I/O address chart

PLC address		Description
Input	I0.0	Start button SB1
	I0.1	Stop button SB2
	I0.2	Magnetic switch B1
Output	Q0.1	Solenoid valve 1YA
	Q0.2	Solenoid valve 2YA

3. Draw the external wiring diagram of PLC

The external wiring diagram of PLC is shown in Fig.2-4-5.

4. Write PLC program

The ladder diagram of PLC is shown in Fig.2-4-6.

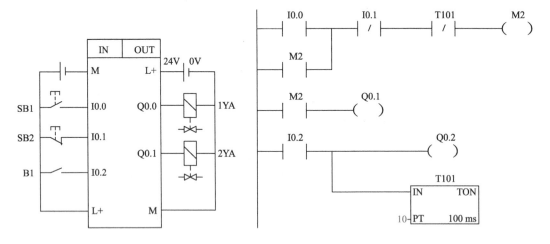

Fig.2-4-5   External wiring diagram of PLC

Fig. 2-4-6   The ladder diagram of PLC

2.4.5 Characteristics and Development Trend of PLC

1. The characteristics of PLC

① High reliability and strong anti-interference ability.

② Strong function, good generality and flexible.

③ Programming is simple, easy to use.
④ Strong expansion capability.
⑤ Short design cycle.

2. The development trend of PLC
① Network.
② Multi-functional.
③ High reliability.
④ Compatibility.
⑤ Miniaturization.

New Words and Phrases

logical ['lɒdʒɪkl]	adj. 符合逻辑的，逻辑上的
programmable ['prəʊgræməbl]	adj. 可设计的，可编程的，可编程序
analog ['ænəlɔːg]	adj. 模拟的
adaptability [ə,dæptə'bɪlətɪ]	n. 适应性，合用性
external [ɪk'stɜːnl]	adj. 外面的，外部的，表面上的，外用的
voltage ['vəʊltɪdʒ]	n. 电压，伏特数
debug [,diː'bʌg]	vt. 排除故障
initialization [ɪ,nɪʃəlaɪ'zeɪʃn]	n. 设定初值，初始化
compatibility [kəm,pætə'bɪlətɪ]	n. 适合，通用性
miniaturization [,mɪnətʃəraɪ'zeɪʃn]	n. 小型化，缩形技术
multifunctional [,mʌlti'fʌŋkʃənl]	adj. 多功能的
relay ['riːleɪ]	n. 继电器
handle ['hændl]	v. 处理，操作
arithmetic operation	算术运算
programmable logic control (PLC)	可编程控制器
central processing unit (CPU)	中央处理单元
anti-interference	抗干扰
sequential function chart (SFC)	顺序功能图
ladder diagram (LAD)	梯形图
statement table (STL)	语句表
function block diagram (FBD)	功能块图
host computer	主机，上位机
input/output unit	输入/输出单元
power unit	电源单元
random access memory (RAM)	随机存储器

read-only memory (ROM) 只读存储器
erasable programmable read only memory (EPROM) 可擦除可编程只读存储器
self diagnosis test 自诊断测试
image register 映像寄存器

Exercises

Ⅰ. Match column A with column B.

A	B
电源单元	host computer
随机存储器	power unit
主机	central processing unit
中央处理单元	miniaturization
小型化	read-only memory (ROM)
只读存储器	random access memory (RAM)

Ⅱ. Mark the following statements with **T** (true) or **F** (false).

(   ) 1. PLC programming is simple and easy to use.
(   ) 2. Only ladder diagram is the programming language of PLC.
(   ) 3. CPU is the core component of PLC.
(   ) 4. PLC has long design cycle.
(   ) 5. Miniaturization is not the development trend of PLC.

Ⅲ. Answer the following questions briefly according to the text.

1. What is the hardware composition of PLC?
2. What is the programming language of PLC?
3. How to work for PLC?
4. What is the development trend of PLC?

 Part B

Reading Material

The Development of PLC

The functionality of the PLC has evolved over the years to include typical relay control, sophisticated motion control, process control, distributed control systems and complex networking. The main differences from other computers are the special input/output arrangements. These connect the PLC to sensors and actuators. PLCs read limit switches, temperature indicators and the positions of complex positioning systems. Some even use machine vision. On the actuator side, PLCs drive any kind of electric motor, pneumatic or hydraulic cylinders or diaphragms, magnetic relays or solenoids. The input/output arrangements may be built into a simple PLC, or the PLC may have external I/O modules attached to a proprietary computer network that plugs into the PLC.

The earliest PLCs expressed all decision making logic in simple ladder logic inspired from the electrical connection diagrams. Today, with the IEC 661131-3 standard, it is now possible to program these devices using structured programming languages, and logic elementary operations.

课文翻译

模块2 控制技术基础

任务1 电气控制

Part A

2.1.1 电气控制的基本概念

电气控制是指利用电气逻辑关系和运算完成对被控对象的控制任务,即为达到某种目的,对被控对象施加所需的操作。通常控制任务根据需求各有差异,需要将所有的被控量和控制装置按照一定的方式连接起来形成一个有机体,即电气控制系统。

电气控制系统包含电气控制元件、电气保护元件、电气执行元件、电气线路、机械传动装置等。在设计过程中,既要满足电气控制系统设计的基本原则,又要充分考虑机械设备的特点,最终形成简单、可靠、经济的控制方案。

2.1.2 电气控制的基本规律

1. 点动控制

点动控制通常由操作者直接操作控制信号(启动按钮),实现电动机的运转。当启动按钮被按下时,电动机运行;当启动按钮被释放时,电动机停止运行。点动控制主要用来实现对生产设备的手动调整、定位、检修处理等。点动控制如图2-1-1所示。

图 2-1-1 点动控制

2. 长动控制

长动控制是通过控制启动信号,实现电动机的运转。当启动信号接通时,电动机运行;当启动信号释放时,由于电路存在自锁功能,所以电动机仍继续运行,只有按下停止按钮,电动机才停止运行。长动控制可以实现电动机的连续运行,长动控制如图2-1-2所示。

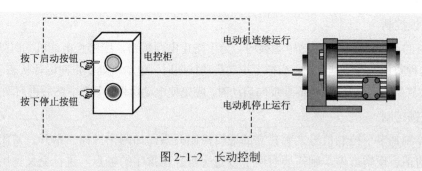

图 2-1-2　长动控制

3. 多点控制

多点控制是操作者可以在不同的地点对电动机实施启动、停止操作。如操作者 A 在电控柜上通过按下启动、停止按钮实现电动机的运行和停止,操作者 B 在生产现场通过按下启动、停止按钮也可以实现电动机的运行和停止。多点控制一般用来控制比较庞大的现场设备或设备工位比较多的生产现场。多点控制如图 2-1-3 所示。

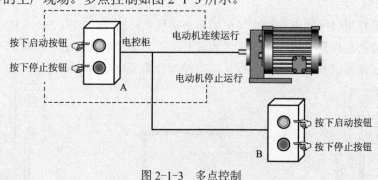

图 2-1-3　多点控制

4. 顺序控制

顺序控制是指多台电动机按照一定的顺序实施启动或停止操作。如图 2-1-4 所示,第一台电动机可以直接启动,第一台电动机启动后允许第二台电动机启动,第二台电动机启动后允许第三台电动机启动;反过来,第三台电动机可以直接停止,第二台电动机在第三台电动机停止后允许停止,第一台电动机在第二台电动机停止后允许停止。

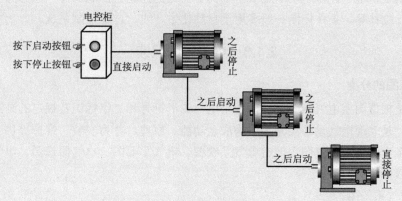

图 2-1-4　多台电动机顺序启动、停止运行

5. 互锁控制

电动机的正、反转需要实施电源换向。为了防止电源短路,应禁止电动机同时处于正、反向供电,这种防护措施就是互锁。当按下正向启动按钮时,电动机处于正向运行状态,此时,反向控制应被屏蔽;反之同理。当需要反向启动时,应先使电动机停止运行,然后再启动反向运行。

6. 联锁控制

联锁控制是多个控制信号之间建立的逻辑控制关系,如两个按钮(常开、常闭)分别控制电动机的正、反转启停,则需要对两个按钮(常开、常闭的触点)进行交叉使用。联锁控制还体现在一组信号对另一组信号的连带关系。目前,使用较多的是按钮联锁控制。

7. 行程控制

行程控制是对生产机械某一运动部件的行程和位置变化情况进行控制。其控制对象是行程开关,它是靠机械外力对行程开关实施碰撞来发出信号的。

8. 时间控制

时间控制是利用时间继电器对生产机械设备实现定时控制。如图2-1-5所示,送料小车在行进过程中碰到行程开关后停止运行,经过某段工艺处理后(延时一段时间)继续向左行驶,直至碰到左侧限位开关后停止运行。时间控制分为通电延时和断电延时。

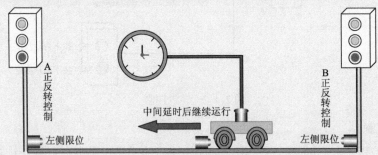

图 2-1-5 送料小车延时控制运行

9. 条件控制

条件控制是指电气控制系统满足某个条件后实施的控制。控制条件一般为温度、压力、流量、液位、位移等。条件控制一般多用于自动化生产线、过程控制系统。

2.1.3　电气图的识读

1. 电气图的分类

电气图是根据国家电气制图标准,使用电气图形符号和文字标识及规定的画法绘制而成的技术图样,是表现电气设备及系统的构成、功能、原理,并为安装、维护提供必要的技术数据依据的工程图样。电气图主要有电气系统图、电气平面图、电气原理图、电气元件布置图、电气接线图等。

(1)电气系统图

电气系统图是用符号或带注释的框,概括表示系统或分系统的基本组成、相互关系及其主要特征的一种简图，如图 2-1-6 所示。

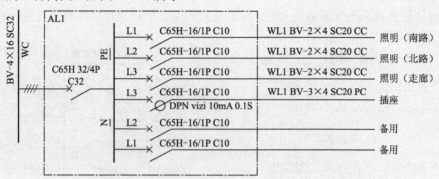

图 2-1-6　照明配电系统图

（2）电气平面图

电气平面图是以建筑平面图为依据，表示设备、装置与管线的安装位置，线路走向，敷设方式等平面布置，而不反映具体形状的简图，如图 2-1-7 所示。通常图样与实际情况有一定的缩放比例关系，是提供安装的主要依据。

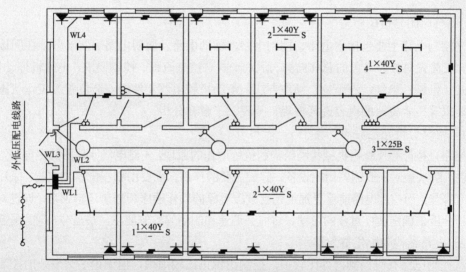

图 2-1-7　照明配电平面图

（3）电气原理图

电气原理图是用图形符号表现电路、设备或系统的工作原理，并按工作顺序排列，体现它们之间的逻辑关系，而不考虑其实际位置的一种简图，如图 2-1-8 所示（图略）。电气原理图分为主电路和控制电路，用以指导安装、接线、调试、使用和维修。

（4）电气元件布置图

电气元件布置图是描述器件的布局和安装位置的一种电气图样，如图 2-1-9 所示（图略）。它包括在电控柜和现场的分布，如电控柜中器件的分布、控制操作盘中器件的分布、器件的间隔和排放顺序、安装方式和定位等。电气元件布置图中，一般标有各元件间距尺寸、安装孔距和进出线的方式。

（5）电气接线图

电气接线图是表示电气设备及元件之间的连接关系，用以进行接线和检查的一种简图，如图 2-1-10 所示。

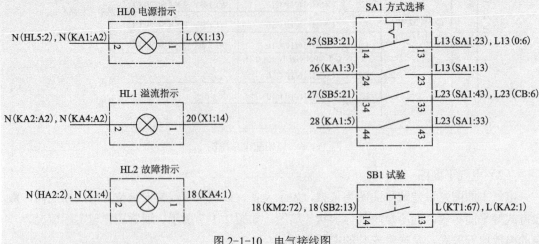

图 2-1-10　电气接线图

2. 电气识图分析技巧

电气控制电路主要包括主电路、控制电路、辅助电路及照明电路等几部分。在阅读、分析之前，应首先对被控对象的总体结构、运动形式、控制流程、控制要求、电动机拖动形式、传动方式（机械、电气、液压）、操作方法、电动机和电气元件的安装位置、工作状态及自动控制要求等技术资料的内容进行搜集、整理、了解和分析。

（1）电路分析先主后辅

主电路直接能够反映出机械传动结构和执行动作的原理，先对电气原理图的主电路进行阅读，可以了解被控对象有哪些用电设备，它们主要的作用，需要由哪些电器来控制，采用的保护措施有哪些。而控制电路能够更加细化出被控对象的动作顺序和控制逻辑关系，通过对控制电路进行分析来确定被控对象的启动、转向、调速和制动等控制要求；最后分析辅助电路。

（2）电路结构分析化整为零

电气控制电路有时控制内容比较多，控制逻辑也比较烦琐，但无论多么复杂的控制电路都是由典型的控制环节组成的。对电气控制电路进行功能划分和逻辑控制关系的梳理，就是常说的化整为零。分析电路时，应从电源侧开始着手，从主令控制开关到接触器、继电器的线圈，由上至下，由左至右逐一进行分析，并注意各个局部控制电路之间的联锁和互锁关系，梳理控制顺序的流程，简洁明了地将控制电路的工作原理及过程表示出来。

（3）集零为整综合分析

经过化整为零的过程之后，初步对电气控制各个环节有了了解，但最终还需要进行集零为整综合分析。从控制电路的整体角度去考虑，清楚各个控制环节之间的内在关系、互锁关系、联锁关系；清楚机械、电气、液压之间协调配合的情况及各种保护环节的设置情况。能够对整个电气控制电路有一个总体的认识，对整个电气工作原理和加工过程的实现有进一步的理解和认识，从而理解和掌握各个电路。

Part B

电气控制系统设计

电气控制系统设计应具有一定的设计原则，应保证被控对象的功能性、安全性、合理性和经济性，要结合控制工艺进行总体规划。电气控制系统设计过程中应满足机械设备对电气控制系统的要求；机电配合趋于合理，实现功能互补；控制系统力求简单、经济、实用；合理选择电气元件和布局方案；尽量保证操作和维修的方便性。

电气控制系统设计主要体现在控制方案设计和控制系统设计两大环节。控制方案设计很重要，它是完成控制系统设计的奠基石，只有好的方案出台，才能有效地进行设计实施。电气控制系统设计应尽量遵循简洁、高效、实用的原则，在完成电气控制系统初期设计后对其再进行优化，最终满足设计要求。

任务 2 液 压 控 制

 **Part A**

2.2.1 液压传动的工作原理

液压传动是利用液体作为传动介质对能量进行传递和控制的一种传动形式，从而实现各种机械的传动和自动控制。相对于机械传动来说，液压传动是一门新技术，利用各种液压元件组成不同功能的液压回路，从而进行能量的传递、转换与控制。

图 2-2-1 所示为液压千斤顶的实物及工作原理图。其结构主要由手动柱塞液压泵（活塞 1、杠杆 2）、液压缸、单向阀、截止阀、油箱等组成。根据帕斯卡定律：在密闭容器内，施加于静止液体上的压力将以等值作用于液体上的各点。其工作过程分析如下：

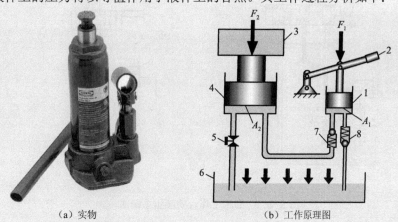

（a）实物　　　　　　　　（b）工作原理图

图 2-2-1　液压千斤顶

1—活塞；2—杠杆；3—负载；4—液压缸；5—截止阀；6—油箱；7、8—单向阀

1．力的传递

液压缸中所产生的液体压力 P_2：$P_2 = \dfrac{F_2}{A_2}$。

作用在液压泵小活塞上的作用力 F_1：$F_1 = PA_1 = P_1A_1 = P_2A_1 = \left(\dfrac{A_1}{A_2}\right) \cdot F_2$

我们可以得到第一个重要的规律：液压传动中工作压力取决于外负载。

2．运动的传递

液压泵排出的液体体积等于进入液压缸的液体体积，则有

$$V_1 = v_1 t \cdot A_1 = v_2 t \cdot A_2 = V_2$$

上式两边同除以运动时间 t 得

$$q_1 = v_1 A_1 = v_2 A_2 = q_2$$

我们可以得到第二个重要的规律：活塞的运动速度只取决于输入流量的大小，而与外负载无关。

从上面的分析中还可以看出，压力和流量是液压传动中两个最基本的参数。

3．液压传动应用实例

图 2-2-2 所示为磨床工作台的液压系统图。其主要由液压泵、液压缸、流量控制阀、方向控制阀（换向阀）、压力控制阀、安全阀、过滤器、油箱等组成。工作时，溢流阀用来调节或限制系统的压力，流量控制阀可调节液压缸运动速度的大小，三位四通换向阀可实现工作台的运动和停止。具体工作油路如下：

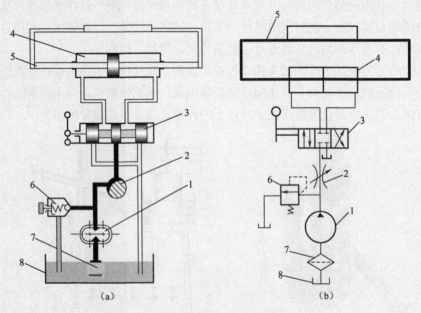

图 2-2-2　磨床工作台的液压系统图

1—液压泵；2—流量控制阀；3—换向阀；4—液压缸；5—工作台；6—溢流阀；7—过滤器；8—油箱

当换向阀左位工作时，液压油经油箱→过滤器→液压泵→节流阀→换向阀（左位）→液

压缸（左腔），推动工作台向右移动。

当换向阀右位工作时，液压油经油箱→过滤器→液压泵→节流阀→换向阀（右位）→液压缸（右腔），推动工作台向左移动。

当换向阀中位工作时，液压油经油箱→过滤器→液压泵→换向阀（中位），由于换向阀是O形中位机能，工作台停止运动。

2.2.2 液压传动系统的组成

任何液压传动系统都可分为以下五个逻辑部分：

1. 能源装置

能源装置可把机械能转换成流体压力能，为液压传动系统提供动力，如液压泵（图 2-2-3，略）。

2. 执行元件

执行元件能把流体的压力能转换成机械能，驱动不同机械机构动作，如用于直线运动的液压缸（图 2-2-4，略）和用于旋转运动的液压马达（图 2-2-5，略）。

3. 控制元件

控制元件可对液压传动系统中流体压力、流量和流动方向进行控制或调节，如溢流阀（图 2-2-6，略）、方向控制阀（图 2-2-7，略）、流量控制阀（图 2-2-8，略）等。

4. 辅助元件

保证系统正常工作所需的上述3种以外的装置，如油箱（图2-2-9，略）、过滤器（图2-2-10，略）、蓄能器（图 2-2-11，略）等。

5. 工作介质

起着能量传递、润滑、防腐、防锈、冷却等作用，如液压油。

为了简化液压传动系统的表示方法，通常采用图形符号来绘制系统原理图。

图 2-2-2（b）就是图 2-2-2（a）所示液压传动系统原理图。

2.2.3 液压传动的特点与应用

1. 液压传动的优点

① 液压传动系统具有在给定功率下，液压装置体积小、重量轻、结构紧凑的特点。
② 液压传动系统工作比较平稳。
③ 液压传动系统能在大范围内实现无级调速。
④ 液压传动系统易于自动化。
⑤ 液压传动系统易于实现过载保护。
⑥ 液压传动系统的设计、制造和使用比较方便。
⑦ 用液压传动系统实现直线运动远比用机械传动简单。

2. 液压传动的缺点

① 存在泄漏现象。

② 工作性能易受温度变化的影响。
③ 液压元件的制造精度要求较高，因而价格较贵。
④ 液压传动出现故障时不易查找原因。

3. 液压传动的应用

液压传动技术在工业中的应用如表 2-2-1 所示。

表 2-2-1 液压传动技术在工业中的应用

行业名称	应用举例
工程机械	挖掘机、装载机、推土机、铲运机等
矿山机械	凿岩机、开掘机、提升机、液压支架等
建筑机械	打桩机、液压千斤顶、平地机等
冶金机械	轧钢机、压力机等
机械制造	机床、数控加工中心、自动装配线、压力机、模锻机等
轻工机械	打包机、注塑机、食品包装机等
汽车工业	高空作业车、汽车起重机、转向器等
水利工程	大坝、闸门、船舶机械、船舵等
农林机械	化肥包装机、联合收割机、拖拉机、农机悬挂系统等

 Part B

液压系统的操作及维护

① 操作者要熟悉液压元件控制机构的操作要领，各个调节手柄的转动方向与所控制的压力或流量大小的变化关系，严防事故发生。

② 工作中应随时注意油位和温升。一般油液的正常工作温度为 30~60 ℃，最高不超过 60 ℃。当异常升温时，应停车检查。冬天气温低时，应使用加热器。

③ 要保持液压油清洁，应定期检查更换。对于新使用的液压设备，使用 3 个月左右就应清洗油箱、更换新油，以后每隔半年至一年进行一次清洗和换油。

④ 注意滤油器的使用情况，滤芯要定期清洗和更换。

⑤ 若设备长期不用，应将各调节手柄全部放松，防止弹簧产生永久变形。

任务 3 气动控制

 Part A

2.3.1 气压传动的工作原理

气压传动与控制技术简称气动技术。气压传动和液压传动实现传动和控制的方法基本相同，它是以压缩空气为工作介质，进行能量或信号传递及控制的一种传动方式，也是控制和驱动各种机械和设备,以实现生产过程自动化的技术。

图 2-3-1 所示为气动剪切机的工作原理图,压缩空气在气动系统中的流动路线如下:空气压缩机 1→冷却器 2→油水分离器 3(降温及初步净化)→储气罐 4(备用)→分水滤气器 5(再次净化)→减压阀 6(调压、稳压)→油雾器 7(润滑元件)→换向阀 9→气缸 10。

气动剪切机的具体工作过程如图 2-3-2 所示(图略),图 2-3-2(a)为剪切前的预备状态,此时行程阀关闭,换向阀上位工作,气缸活塞向下移动,剪刃张开。

图 2-3-2(b)为剪切时的状态,当工件被送入剪切机规定位置时,工件压下行程阀,行程阀与大气接通,换向阀阀芯下移,下位工作,气缸活塞带动剪刃快速向上运动,将工件切断。

当工件被切下后,即与行程阀脱开,行程阀复位,换向阀阀芯上移复位。气缸活塞带动剪刃向下运动,系统又恢复到图 2-3-2(a)预备状态,为第二次进料剪切做准备。

从上述实例可以得出结论:气压传动是以压缩空气作为工作介质,依靠气体在密闭容积中的压力能来传递运动和动力,其实质是机械能与气体压力能之间的能量转换,即"机械能→气体压力能→机械能"。

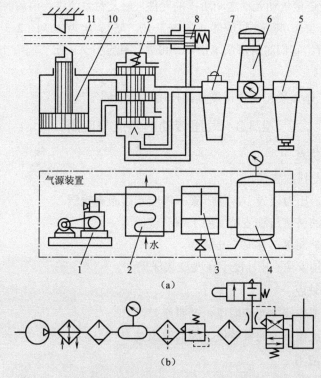

图 2-3-1 气动剪切机的工作原理图

1—空气压缩机;2—冷却器;3—油水分离器;4—储气罐;5—分水滤气器;
6—减压阀;7—油雾器;8—行程阀;9—换向阀;10—气缸;11—工件

2.3.2 气压传动系统的组成

从气动剪切机的工作原理图可知,一个完整的气动系统有以下组成部分:

1. 气源装置

如图 2-3-3 所示（图略），气源装置是向气压传动系统提供压缩空气（具有一定的压力和流量，并具有一定的净化程度）的装置，包括将机械能转化为气体压力能的空气压缩机，净化、储存压缩空气的装置和设备，传输压缩空气的管道系统等。

2. 气动执行元件

气动执行元件能将空气压缩机中气体的压力能转化为机械能，驱动各种机械装置动作。包括做直线运动的气缸（图 2-3-4，略）和做回转运动的气马达（图 2-3-5，略）。

3. 气动控制元件

气动控制元件能对气动系统中压缩空气的压力、流量和流动方向进行控制和调节。如气动减压阀（图 2-3-6，略）、节流阀（图 2-3-7，略）、换向阀（图 2-3-8，略）等。这些元件的不同组合，可以实现气动系统的不同功能。

4. 气动辅助元件

气动辅助元件能连接气动元件之间所需的元件，以及对系统进行消声、冷却、测量等。如管道、压力表（图 2-3-9，略）、过滤器（图 2-3-10，略）、消声器（图 2-3-11，略）、油雾器（图 2-3-12，略）等。它们对保持系统正常、可靠、稳定持久的工作起着十分重要的作用。

5. 工作介质

在气动系统中工作介质起到传送能量的作用，如压缩空气。

2.3.3　气压传动的特点与应用

1. 气压传动的优点

① 空气的获得与排放方便。

② 气体黏性低，压力损失小，便于集中供应和远距离输送。

③ 对气动元件的材料与制造精度要求相对较低。

④ 气动系统维护简单，管道不易堵塞。

⑤ 气动系统使用安全，并且便于实现过载保护。

2. 气压传动的缺点

① 气动系统的工作稳定性较液压传动系统差。

② 气压传动的工作压力通常较低，因而输出功率较小。

3. 气动技术的应用

由于气动的工作介质为压缩空气，其具有防火、防爆、防电磁干扰、抗振动、无污染、结构简单和工作可靠等特点，所以气动技术与液压、机械、电子技术相结合的方式，已发展成为实现生产过程自动化的一个重要手段。

气动技术现在被广泛应用于机械、电子、轻工、纺织、食品、医药、包装、冶金、石化、航空和交通运输等各个工业部门，如气动机械手、组合机床、加工中心、生产自动线、自动检测和实验装置等。气动技术在提高生产效率、自动化程度、产品质量、工作可靠性等方面显示出极大的优越性。

Part B

气动系统的故障诊断方法

1. 经验法

经验法可根据实际经验和简单的仪表进行故障的诊断和排除,可按照"望、闻、问、切"的中医诊断模式进行。

2. 推理分析法

推理分析法是根据逻辑推理寻找出故障的真实原因。

(1) 仪表分析法

利用仪器仪表,检测系统或元件的技术参数是否符合气动系统要求。

(2) 部分停止法

暂时停止气动系统某部分的工作,观察对故障征兆的影响。

(3) 试探反证法

试探性地改变气动系统中部分工作条件,观察对故障征兆的影响。

(4) 比较法

用标准的或合格的元件代替系统中相同的元件,通过工作状况的对比,来判断被更换的元件是否失效。

任务 4 PLC 控 制

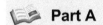 **Part A**

2.4.1 PLC 的定义

可编程控制器(PLC)是一种数字运算操作的电子系统,专为在工业环境下应用而设计。它采用了可编程序的存储器,用来在其内部存储执行逻辑运算、顺序控制、定时、计数和算术运算等操作的指令,并通过数字式和模拟式的输入和输出,控制各种类型机械的生产过程。

PLC 具有面向过程、面向用户、适应工业环境、操作方便、可靠性高的特点。它的控制技术代表着当前程序控制的先进水平,并且已经成为基本自动控制系统装置。

2.4.2 PLC 的组成

可编程控制器是一种通用工业控制计算机,由硬件和软件两部分组成。

1. PLC 的硬件组成

PLC 型号种类繁多,但结构和工作原理基本相同。PLC 通常由五部分组成:中央处理单元(CPU)、存储器、输入/输出单元、电源单元、编程器,如图 2-4-1 所示。

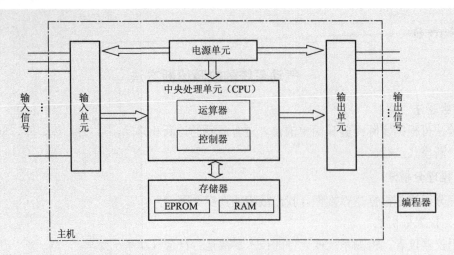

图 2-4-1 PLC 的硬件组成

（1）中央处理单元（CPU）

CPU 是 PLC 的核心部件，类似于人的大脑，包括运算器、控制器。

（2）存储器

存储器用于存储数据或程序，它包括随机存储器（RAM）、只读存储器（ROM）、可擦除可编程只读存储器（EPROM）。

（3）输入/输出单元

输入/输出单元是 PLC 与 I/O 设备或其他外围设备相互联系的通道，能输入/输出各种操作电平和驱动信号。

（4）电源单元

电源单元是 PLC 为其内部电路进行供电的开关式稳压电源。

（5）编程器

编程器是 PLC 的主要外围设备，用以实现用户程序的编制、编辑、调试和监视，还可以通过其键盘来调用和显示 PLC 内部器件的状态和系统参数。因此，它也是系统运行和故障分析的一个主要工具。

2. PLC 的编程语言

（1）顺序功能图（SFC）

（2）梯形图（LAD）

（3）语句表（STL）

（4）功能块图（FBD）

（5）高级语言

2.4.3　PLC 的工作过程

如图 2-4-2 所示，PLC 的工作是周而复始地循环扫描的过程。整个扫描工作过程包括内部处理、通信处理、输入扫描、用户程序执行、输出处理 5 个阶段。扫描完整个过程所需的时间称为扫描周期,扫描周期的长短与用户程序的长短和扫描速度有关。

1. 内部处理

内部处理包括系统初始化和自诊断测试。

2. 通信处理

主要完成 PLC 之间、PLC 与上位机或终端设备之间的信息交换。

3. 输入扫描

在用户程序执行之前,PLC 以扫描方式顺序读入所有输入端子的状态,并存入输入映像寄存器中,此时,输入映像寄存器中的内容被刷新。

4. 用户程序执行

用户程序执行时,CPU 从第一条指令开始,以扫描方式按顺序执行程序指令,直到最后一条指令结束。

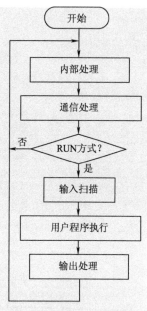

图 2-4-2　PLC 的工作过程

5. 输出处理

执行完所有程序指令后,PLC 将元件映像寄存器中所有的输出继电器的状态以批处理的方式存到输出锁存寄存器中,再经过一定的输出方式,驱动外部负载。

2.4.4　PLC 应用实例

图 2-4-3 所示为雨伞试验机工作原理图,它可以连续模拟自动开伞动作,完成无故障连续开伞次数试验。

其工作要求为:

常态下,气缸 1 活塞杆缩回,气缸 2 活塞杆伸出。

工作时,按下启动按钮→1YA"+"→气缸 2 活塞杆缩回(合上雨伞)→压下磁性开关 B1→2YA"+"→气缸 1 活塞杆伸出(压下开伞按钮)→延时 1s 后→1YA"-"→气缸 2 活塞杆伸出(打开雨伞)→2YA"-"→气缸 1 活塞杆缩回(松开开伞按钮),完成一次开伞试验。

该气动系统如何通过 PLC 来控制呢?以下是 PLC 系统的设计过程。

1. 绘制雨伞试验机系统的电气控制图

根据雨伞试验机工作过程绘制出如图 2-4-4 所示的电气控制图(图略),其控制过程如下:

按下启动按钮 SB1→继电器 KZ"+"→电磁阀 1YA"+"(气缸 2 活塞杆缩回);→磁性开关 B1 接通→电磁阀 2YA"+"(气缸 1 活塞杆伸出)→时间继电器 KT"+"(延时 1 s)→电磁阀 1YA"-"(气缸 2 活塞杆伸出)→磁性开关 B1 断开→电磁阀 2YA"-"(气缸 1 活塞杆缩回)。

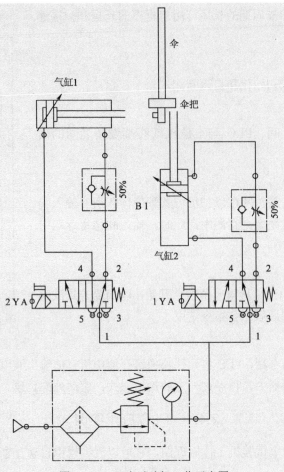

图 2-4-3 雨伞试验机工作示意图

2. 分配 I/O 地址

根据 PLC 系统的要求,I/O 地址分配如表 2-4-1 所示。

表 2-4-1 I/O 地址分配

	PLC 地址	说　明
输入	I0.0	启动按钮 SB1
	I0.1	停止按钮 SB2
	I0.2	磁性开关 B1
输出	Q0.1	电磁阀 1YA
	Q0.2	电磁阀 2YA

3. 绘制 PLC 的外部接线图

PLC 的外部接线图如图 2-4-5 所示(图略)。

4. 编写 PLC 程序

PLC 的梯形图如图 2-4-6 所示(图略)。

2.4.5　PLC 的特点及发展趋势

1. PLC 的特点
① 可靠性高、抗干扰能力强。
② 功能强、通用性好、使用灵活。
③ 编程简单、易于使用。
④ 扩展能力强。
⑤ 设计周期短。

2. PLC 的发展趋势
① 网络化。
② 多功能。
③ 高可靠性。
④ 兼容性。
⑤ 小型化。

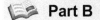 **Part B**

PLC 的发展

可编程控制器经过数年的发展，其功能包括典型的继电器控制，复杂的运动控制，过程控制，分布控制系统和复杂的网络技术。与其他计算机比较，最大的区别是 PLC 有专用的输入/输出装置。这些装置将 PLC 与传感器和执行机构相连接。PLC 能读取限位开关、温度指示器和复杂定位系统的位置信息，有些甚至使用了机器视觉。执行机构方面，PLC 能驱动任何一种电动机、气压或液压装置、光控装置、电磁继电器或螺线管。输入/输出装置可能安装在 PLC 内部，或者通过外部 I/O 模块与专用的计算机网络相连。

最早的 PLC 采用简单的梯形图来描述其逻辑程序，梯形图源自于电气连线图。如今，依照国际电工委员会的 61131-3 标准，现在 PLC 可以使用结构化的编程语言和逻辑初等变换来进行编程。

Module 3 Application of Mechanotronics Device

Task 1 CNC Machine

 **Part A**

Text

3.1.1 The General of CNC Machine

1. CNC machine working process

Firstly, the information for machining workpiece required to process is recorded by the program that stored in some media when CNC machines work, and then the program is input to the numerical control device. Processing by the CNC device, command and control signals are issued to servo system that drive machine, coordinate machine movement and made it produce a series of machine movement such as the main motion and feed motion, so complete workpieces machining.

2. CNC machine advantages

CNC machines have many advantages over conventional machines, some of them are:

① There is a possibility of performing multiple operations on the same machine in one setup.

② Because of the possibility of simultaneous multi-axis movement, special profile tools are not necessary to cut unusual shapes.

③ The scrap rate is significantly reduced because of the precision of the CNC machine and lesser operator impact.

④ Production is significantly increased.

3. CNC machine disadvantages

CNC machines also have some disadvantages, some of them are:

① They are quite expensive.

② They have to be programmed, set up, operated, and maintained by highly skilled personnel.

3.1.2 CNC Lathe

1. The general of CNC lathe

CNC lathe is computer numerical control lathe. The most essential operations on the CNC lathe are facing, cylindrical turning, grooving, parting, drilling, boring and threading.

2. CNC lathe's components (Fig.3-1-1)

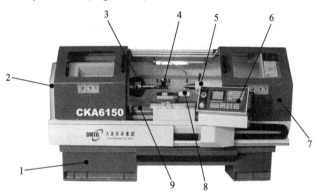

Fig.3-1-1 Components of the CNC lathe
1-bed; 2-headstock; 3-three jaw chuck; 4-turret; 5-tailstock;
6-operation panel; 7-protective shield; 8-turret motor; 9-guide way

The bed mounted on the base, which carries all main components of the lathe and ensures their accurate relative positions during machining.

The headstock locates on the left of the bed. The main function of the headstock is supporting the spindle and accommodating spindle transmissions for the primary motion of the lathe.

Tool post slides include a longitudinal slide that locates on bed ways to perform Z-direction motion, and a cross slide locates on longitudinal slide for X-direction motion. It realizes longitudinal and transverse feeding motion of the cutting tools that are carried.

The turret post is mounted on the slide to hold cutting tools, which can be automatically selected as required for certain machining purpose.

The tailstock is mounted on bed ways and can be moved longitudinally along the ways. It holds a center to provide an assistant support for the workpiece during machining. It can also carry tools (e.g. drills, reamers) for holes making.

The protective shield is mounted on the base to protect operators and the workshop from chip.

Hydraulic transmission provides some assistant motions, mainly, spindle gear shift, tailstock moving, and automatic clamping.

The control system consists of a CNC system (including a CNC device, a servo system and a PLC) and an electrical control system. It realizes automatic machine tool control.

3. CNC lathe's classification

(1) Classification by the main spindle's position

According to the main spindle's position CNC lathes are classified as horizontal (Fig.3-1-1) and vertical (Fig.3-1-2) CNC lathe. Vertical CNC lathes are used to turn large diameter disc-type workpieces. Horizontal CNC lathes are used to machine revolutionary surfaces on shafts, bushes, plates, e.g. inner or outer cylindrical surfaces, conical surfaces, threads.

(2) Classification by the feature of the control system

According to the feature of the control system CNC lathes are classified as economy CNC lathes, full-function CNC lathes, turning centers and FMC lathe.

Fig.3-1-2 Vertical CNC lathe

The design of economy CNC lathes (Fig. 3-1-3) is usually a modification based on that of conventional lathes. They are equipped with open loop servo-mechanisms, and use single chip microcomputers as their CNC devices. Economy CNC lathes have simple constructions and are less expensive, compared to other CNC lathes, they lack of the functions of automatic tool tip radius compensation and constant linear velocity control.

Full-function CNC lathes are known as CNC lathes or standard CNC lathes (Fig. 3-1-4), representing standard control system. It has a high resolution CRT. Graphics simulation, tool compensation, communication/network and multi-axis controllability are the standard functions of the system. Full-function CNC lathes have closed loop or semi-closed loop controllers for accuracy. Besides, high rigidity and high efficiency are highlighted.

Fig. 3-1-3 Economy CNC lathe   Fig. 3-1-4 Full-function CNC lathe

Turning centers (Fig. 3-1-5) are designed to increase productivity. Besides the functions of full-function CNC lathes, a turning center is equipped with a tool magazine, ATC, indexing device, milling unit and even robot arms. Therefore, different machining process such as turning, milling, drilling, reaming screw tapping, etc. can be performed by clamping once. Although turning centers have better efficiency and automation than full-function CNC lathes, they are costly.

An FMC lathe (Fig. 3-1-6) is actually a flexible manufacturing cell consists of a CNC lathe and a robot. It is capable to perform automatic transfer, clamping and machining of a workpiece, preparations and adjusting are also automatic.

Fig. 3-1-5　Turning center

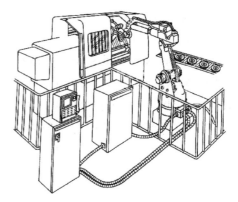

Fig. 3-1-6　An FMC lathe

3.1.3　CNC Machining Center

1. The overview of CNC machining center

Machining centers have been defined as multifunction CNC machines with automatic tool-changing capabilities and rotating tool magazine. Increased productivity and versatility are major advantages of machining centers. They have the ability to perform drilling, turning, reaming, boring, milling, contouring and threading operations on a single machine. Most workpieces can be completed on a single machining center, often with one setup.

2. The components of CNC machining center

The main parts of CNC machining centers are bed, saddle, column, table, servo system, spindle, tool changer and the machine control unit (Fig.3-1-7).

The bed is usually made of high-quality cast iron which provides for a rigid machine the capability of performing heavy-duty machining and maintaining high precision. The bed supports all the components.

The table, which is mounted on the bed, provides the machining center with the X-axis linear movement.

The saddle, which is mounted on the

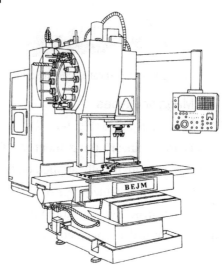

Fig.3-1-7　The components of the CNC machining center

hardened and ground bed ways, provides the machining center with the Y-axis linear movement.

The column is mounted to the saddle. The column provides the machining center with the Z-axis linear movement.

The servo system, which consists of servo drive motors, ball screws, and position feedback encoders, provides fast, accurate movement and positioning of the X, Y or Z axis.

The tool changer is capable of storing a number of preset tools which can be automatically called for use by the workpiece program. The tool change time is usually only 3~5 s.

Machine control unit (MCU) is a computer used to store and process the CNC programs entered. MCU varies according to manufacturer's specifications; new MCU is becoming more sophisticated, making machine tools more reliable and the entire machining operations less dependent on human skills.

3. CNC machining center's classification

Machining centers are classified as vertical (Fig.3-1-8), horizontal (Fig.3-1-9), universal (Fig.3-1-10) and gantry (Fig.3-1-11). Vertical machining centers can be widely accepted and used, primarily for flat parts. Horizontal machining centers are also widely accepted and used, particularly with large, boxy and heavy parts. Universal machining centers are equipped with both vertical and horizontal spindles. With a variety of features, they are capable of machining all surfaces of a workpiece. Gantry machining centers are suitable for machining large size complex parts, especially with high accuracy requirements, e.g. large turbine blades.

Fig.3-1-8 Vertical machining center

Fig.3-1-9 Horizontal machining center

4. Magazine types

(1) Linear magazine (Fig.3-1-12)

Tools are in linear arrangement. As the simplest magazine, it can only hold 6~8 tools. Linear magazines are usually applied on CNC lathe, and sometimes CNC driller.

(2) Circular magazine (Fig.3-1-13)

A circular magazine can store up to 6~60 tools, and can be in many forms. The tools can layout such as radial arrangement, axial arrangement and conical arrangement.

Fig.3-1-10　Universal machining center

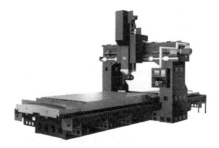

Fig.3-1-11　Gantry machining center

Fig.3-1-12　Linear magazine

Fig.3-1-13　Circular magazine

(3) Chain magazine (Fig.3-1-14)

Chain magazines are widely used. Chain magazine is capable to hold 30~120 tools.

(4) Pigeonhole magazine (Fig.3-1-15)

The pigeonhole magazine has large capacity. The entire magazine can be exchanged with another that has been prepared. The compact construction of the magazine occupies less space. However, the motions of tool selection and transfer are complex, therefore, pigeonhole magazines are commonly used in FMS rather than a single MC.

Fig.3-1-14　Chain magazine

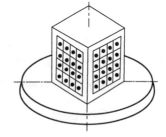

Fig.3-1-15　Pigeonhole magazine

3.1.4　Safety Notes for CNC Machine Operation

Safety is always a major concern in a metal-cutting operation. CNC equipment is automated and very fast, and consequently it is a source of hazards. The hazards have to be located and the operator must be on guard of them in order to prevent injuries and damage to the equipment. Main potential hazards includes: rotating parts, such as the spindle, the tool in the spindle, chuck, part in the chuck, and the turret with the tools and rotating clamping devices; movable parts, such as the machining center table, lathe slides, tailstock center, and turret; errors in the program such as improper use of the G00 code in conjunction with the wrong coordinate value, which can generate an unexpected rapid motion; an error in setting or changing the offset value, which can result in a collision of the tool with the part or the machine; and a hazardous action of the machine caused by unqualified changes in a proven program. To minimize or avoid hazards, try the following preventive actions:

① Keep all of the original covers on the machine as supplied by the machine tool builder.

② Wear safety glasses, gloves, and proper clothing and shoes.

③ Do not attempt to run the machine before you are familiar with its control.

④ Before running the program, make sure that the part is clamped properly.

⑤ When proving a program, follow these procedures:

a. Run the program using the machine lock function to check the program for errors in syntax and geometry.

b. Slow down rapid motions using the "RAPID OVERRIDE" switch or dry run the program.

c. Use a single-block execution to confirm each line in the program before executing it.

d. While the tool is cutting, slow down the feed rate using the "FEED OVERRIDE" switch to prevent excessive cutting conditions.

⑥ Do not handle chips by hands and do not use chip hooks to break long curled chips. Stop the machine if you need to properly clean chips.

⑦ Keep tool overhang as short as possible, since it can be a source of vibration that can break the insert.

⑧ Stop the machine when changing the tools, indexing inserts, or removing chips.

⑨ Do not make changes in the program if yours supervisor has prohibited you from doing so.

⑩ If you have any safety-related concerns, notify your instructor or supervisor immediately.

New Words and Phrases

shaft [ʃɑːft]	n. 柄，轴，矛，箭
bush [buʃ]	n. 灌木（丛），（金属）衬套
graphics ['græfɪks]	n. [测]制图学，制图法，图表算法
simulation [ˌsɪmjuˈleɪʃn]	n. 模仿，模拟
flexible ['fleksəbl]	adj. 灵活的，柔韧的，易弯曲的
clamp [klæmp]	vt.& vi. 夹紧，夹住，锁住
rigorous ['rɪgərəs]	adj. 严格的，严密的，缜密的
headstock ['hedstɒk]	n. 主轴箱，车头，头座

longitudinal [ˌlɒŋgɪˈtjuːdɪnl]	adj. 纵向的，纵的，经度的
combination [ˌkɒmbɪˈneɪʃn]	n. 结合，联合体
turret [ˈtʌrət]	n. 炮塔，转塔，塔楼
tailstock [ˈteɪlstɒk]	n. 车尾，床尾，尾架，尾座，顶针座
layout [ˈleɪaʊt]	n. 布局，安排，设计，布置图，规划图
gantry [ˈgæntri]	n. 构台，桶架
universal [ˌjuːnɪˈvɜːsl]	adj. 普遍的，一般的，通用的，万能的
turbine [ˈtɜːbaɪn]	n. 汽轮机，涡轮机，透平机
blade [bleɪd]	n. 桨叶，刀片，剑
conical [ˈkɒnɪkl]	adj. 圆锥（形）的
pigeonhole [ˈpɪdʒɪnhəʊl]	n. 小房间，文件架上的小间隔
disc-type	盘式
full-function CNC lathe	全功能型数控车床
turning center	车削中心
turret tool	转塔刀架
ball screw	滚珠丝杠
tool magazine	刀库
gantry machining center	龙门加工中心
pigeonhole magazine	箱式刀库

Exercises

Ⅰ. Match column A with column B.

A	B
箱式刀库	turning center
龙门加工中心	pigeonhole magazine
进给倍率	ball screw
车削中心	rotating tool magazine
滚珠丝杠	gantry machining center
旋转刀库	feed override

Ⅱ. Mark the following statements with T (true) or F (false).

(　　) 1. The CNC lathes are classified as horizontal, vertical, and universal CNC lathe.

(　　) 2. Economy CNC lathes have closed loop or semi-closed loop controllers.

(　　) 3. Hydraulic transmission provides some assistant motions, mainly, spindle gear shift, tailstock moving, and automatic clamping.

(　　) 4. The pigeonhole magazine has large capacity.

(　　) 5. Gantry machining centers are equipped with both vertical and horizontal spindles.

III. Answer the following questions briefly according to the text.

1. What's definition of the CNC lathes?

2. What's the function of the tailstock?

3. Which four kinds are machining center divided into?

4. Is it true that the machining center has the ability to perform drilling, reaming and boring operations on a single machine?

5. How much tools are circular magazine stored?

Part B

Reading Material

Classification of CNC Machine

According to tool motions, CNC machines are classified as point-to-point (PTP) and contouring(or continues path). Point-to-point control systems position the tool from one point to another. The simplest example of such a system is a drilling machine. The drill path and its feed while traveling from one point to the next point are assumed to be unimportant. The path from the starting point to the final position is not controlled. A contouring control system is able to regulate the rate of the table (or spindle) for at least two axes simultaneously. This control requires an elaborate control and drive system.

According to machine control loops, CNC machines are classified as open-loop and closed-loop. The open-loop control system does not provide positioning feedback to the control unit. The advantage of the open-loop control system is that it is less expensive. The disadvantage is the difficulty of detecting a positioning error. Closed-loop systems are very accurate. They use AC, DC or hydraulic servo motors.

Task 2 CNC EDM Machine

Part A

Text

3.2.1 CNC Sinker EDM Machine

1. The basic principal of CNC sinker EDM (Fig.3-2-1)

Electrical discharge machining (EDM), spark machining, as it is so called, is based on the eroding effect of an electric spark on both the electrodes used to produce it. Impulse power is linked to the tool electrode and the workpiece. Using automatic feed adjustment device makes the tool electrode and the workpiece all along keep a small discharge gap. By the action of impulse electric current, spark discharging is produced. There is big electric current density at the convex place on the workpiece and the electrode surface and partial high temperature is produced, so the convex place will be fist melted, gasified to form minute concave pit, which will form a new convex place where metal is corroded when impulse is discharging next time. The melted metal power is scattered into the working liquid to be taken away and filtered. The constant impulse discharge will replicate the tool shape of the tool electrode on the workpiece to carry out the shaped machining.

2. The components of CNC sinker EDM machine

CNC sinker EDM machine (Fig.3-2-2) is a kind of high-precision automatic machining machine. It is composed of machine itself, impulse power, automatic feed adjustment device, CNC system, working liquid cycle system and so on. Machine spindle is equipped with tool electrode (positive electrode); workpiece (negative electrode) is clamped on the worktable. Under the CNC control, Z-direction servo motor drives the spindle to move up and down through ballscrew and keeps steady discharge gap between tool electrode and workpiece to carry out arc machining process. With servo motor, through ballscrew, worktable can perform X, Y-direction feed movement to finish the defined path machining.

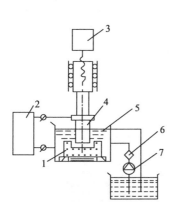

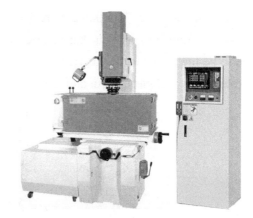

Fig. 3-2-1 The basic principle of CNC sinker EDM

1-workpiece; 2-pulse power; 3-servo mechanism; 4-electrode;
5-dielectric fluid; 6- dielectric fluid filter; 7- pump

Fig.3-2-2 CNC sinker EDM machine

3. Characteristics and application of CNC Sinker EDM

(1) Characteristics of sinker EDM

① Some of the advantages of CNC sinker EDM include:

a. Complex shapes that would otherwise be difficult to produce with conventional cutting tools.

b. Extremely hard material to very close tolerances.

c. Very small workpieces where conventional cutting tools may damage the workpieces from excess cutting tools pressure.

d. There is no direct contact between tool and workpiece. Therefore delicate sections and weak materials can be machined without any distortion.

② Some of the disadvantages of CNC sinker EDM include:

a. The slow rate of material removal.

b. The additional time and cost used for creating electrodes for sinker EDM.

c. Reproducing sharp corners on the workpiece is difficult due to electrode wear.

d. Power consumption is very high.

(2) CNC sinker EDM applications

Because CNC sinker EDM is able to create a wide variety of difficult shapes, it has become popular for many different applications. CNC sinker EDM has found widespread use not only in the manufacture of mold making, but also in aerospace applications and the production of small holes and micro holes.

3.2.2 CNC WEDM Machine

1. The basic principal of CNC WEDM (Fig.3-2-3)

The basic principal of CNC WEDM is that the continuously moving thin metal wire is taken as tool electrode. The pulse current is linked to the metal wire and the workpiece, and then with the impulse spark discharging between the metal wire and the workpiece, the metal is made to be melted or to be gasified. During the relative movement of the electrode wire and workpiece, the workpiece is cut and shaped. CNC WEDM is also called wire cut.

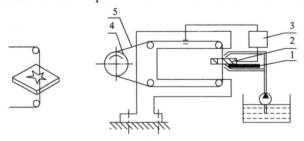

Fig. 3-2-3   The basic principal of CNC WEDM

1-insulation base plate; 2-workpiece; 3-impulse power; 4-wire spool; 5- electrode wire

2. The classification of CNC WEDM machine

CNC WEDM can be classed into the following two kinds according to the electrode wire movement speed.

(1) CNC WEDM-HS (Fig.3-2-4)

The electrode wire of this kind of machine carries out high-speed and reciprocating movement. So it is also called reciprocating or rapid wire cutting machine.

(2) CNC WEDM-LS (Fig.3-2-5)

The electrode wire of this kind of machine carries out low-speed and single movement. So it is also called single or slow wire cutting machine.

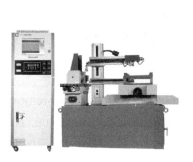

Fig.3-2-4　CNC WEDM-HS　　　　Fig.3-2-5　CNC WEDM-LS

3. The components of CNC WEDM Machine

CNC WEDM is mainly composed of machine itself, impulse power, working liquid, CNC system and machine accessories and so on (Fig.3-2-6).

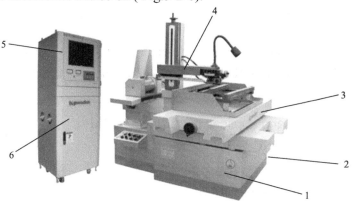

Fig.3-2-6　The components of CNC WEDM
1-bed; 2- working liquid; 3-coordinated worktable;
4-wire rest; 5-CNC system; 6- impulse power

(1) Machine itself

① Bed. Bed is a support of the coordinate worktable, string wire device and wire rest. It should have enough strength and rigidity. Power and working liquid box may be mounted inside the bed.

② Coordinated worktable. The coordinated worktable is used to mount workpieces and in performs the predetermined relative movement with respect to electrode wire according to control requirement. It involves carrier, way and drive device and so on.

③ Wire rest. The effect is that the wire rest can support and guide the electrode wire by means of two guide pulleys on the wire rest and can make some parts of the electrode wire and worktables keep certain angle to perform taper cutting. Double coordinate gang wire has added

U and V two drive-motors to carry out taper cutting by program control.

(2) Impulse power

The effect is that the impulse power change power frequency alternating current into single impulse current at a certain frequency in order to provide electric energy which workpiece and electrode wire requires when they discharging to corrode metal. Its character directly affects machining speed, surface quality, machining precision and the loss of electrode wire.

(3) Working liquid

In the machining, working liquid plays roles of insulation, flushing, chip-removal, cooling and it can affect cutting speed, surface roughness, machining precision and so on. Therefore the requirements for the working liquid are as follows.

① Having insulating property.

② Have very strong permeability.

③ Cooling property should be good.

④ Not have environmental pollution and should not pose dangers to human beings.

⑤ Low price and steadiness are good.

(4) CNC system

The main effect is that CNC system auto-control the relative moving path of the electrode wire and workpiece as well as feed speed to carry out automatically machining. Its main function is path control and machining control.

4. Characteristics and application of CNC WEDM

(1) Characteristics of CNC WEDM

① Some of the advantages of CNC WEDM:

a. Material of any hardness can be cut.

b. High accuracy and good surface finish are possible.

c. No cutting forces involved.

② Some of the disadvantages of CNC WEDM:

a. Limited to electrically conductive materials.

b. Slow process, particularly if good surface finish and high accuracy are required.

(2) CNC WEDM applications

CNC WEDM is one of the most common types of electrical discharge machining, which uses the most accurate types of machining equipment to manufacture hard metals and other materials that are difficult to machine cleanly with conventional mechanical-cutting methods. Industries that utilize CNC WEDM machining technology include electronics, tool and die, aerospace, robotics and medical.

Module 3 Application of Mechanotronics Device

New Words and Phrases

discharge [dɪsˈtʃɑːdʒ]	vt.& vi. 放出，流出，发射
electrical [ɪˈlektrɪkl]	adj. 用电的，与电有关的，电学的
removal [rɪˈmuːvl]	n. 除去，搬迁，移走
impulse [ˈɪmpʌls]	n. [电子]脉冲
electrode [ɪˈlektrəʊd]	n. 电极，电焊条
convex [ˈkɒnveks]	adj. 凸形，凸的，凸面的
gasified [ˈɡæsɪfaɪd]	adj. 气化的 v. 使成为气体，使气化 (gasify 的过去式和过去分词)
reciprocate [rɪˈsɪprəkeɪt]	vi. 回报，往复运动
accessory [əkˈsesəri]	n. 附件
predetermine [ˌpriːdɪˈtɜːmɪn]	vt.& vi. 预先裁定，注定
pulley [ˈpʊli]	n. 带轮，滑轮（组），滑车
gang [ɡæŋ]	n.（工具，机械等的）一套
taper [ˈteɪpə(r)]	vt.& vi. 逐渐变细，逐渐减弱
insulation [ˌɪnsjuˈleɪʃn]	n. 绝缘，隔离，绝缘或隔热的材料
permeability [ˌpɜːmɪəˈbɪləti]	n. 渗透性，磁导率，可渗透性
electrical discharge machining	电火花加工
negative electrode	负极
discharge gap	放电间隙
sharp corner	尖角
electrode wire	电极丝
wire cut	线切割
wire rest	丝架
taper cutting	锥度切割
reciprocating movement	往复运动

Exercises

Ⅰ. Match column A with column B.

A	B
往复运动	wire cut
线切割	sharp corner
负极	reciprocating movement
放电间隙	electrode wire
电极丝	discharge gap
尖角	negative electrode

Ⅱ. **Mark the following statements with T (true) or F (false).**

(　　) 1. EDM is based on the eroding effect of an electric spark on both the workpiece used to produce it.

(　　) 2. Impulse power is linked to the tool electrode and the workpiece.

(　　) 3. CNC sinker EDM machine is a kind of low-precision automatic machining machine.

(　　) 4. WEDM is that the continuously moving plastic wire is taken as tool electrode.

(　　) 5. Bed is a support of the coordinate worktable, string wire device and wire rest.

Ⅲ. **Answer the following questions briefly according to the text.**

1. What is the basic principle of CNC sinker EDM?

2. What are the components of CNC sinker EDM?

3. What are the applications of CNC WEDM?

4. Which can be classified into two kinds of CNC WEDM according to the electrode movement speed?

 Part B

Reading Material

The Choosing of the Electrical Criteria

The users must choose the electrical criteria such as current, pulse width, pulse interval and so on according to the working requirements.

1. The rough machining

In order to obtain faster working speed, the users should choose a large pulse width and high current. The current selection should consider the electrode size in order to avoid too much current per unit area. Considering the rate of working, the pulse interval should be chosen something as small as possible. But too small pulse interval is deterioration to the working condition; indirectly will increase the electrode wear. In order to obtain a smaller electrode wear, the negative of working should be chosen, i.e. the workpiece will connect with negative, tool electrode will connect with the positive electrode.

2. The finishing machining

The primary purpose of the finishing is getting a good surface roughness and dimensional accuracy. The pulse width and current should be chosen smaller. Because of the poor chip removal condition, the pulse interval should be chosen larger. Lift the tool frequently but the height is low. The pulse width should be chosen less than 80 μs and ensure the discharge stable.

Task 3　Robot

 Part A

Text

3.3.1　Robot Overview

Robot technology is an interdisciplinary comprehensive technology, which involves mechanics, machinery, electrical hydraulic technology, automatic control technology, sensor technology, computer science and other scientific fields. Currently the robots of Mitsubishi, KUKA, ABB and other companies are applied in industry.

1. Definition of robot

The Robot Institute of America has developed the following definition: A robot is a programmable, multi-function device designed to move material, parts, tools, or special devices through variable programmed motions for the performance of a variety of tasks.

2. Development of robot

The first robot was invented by Devol and Engelberger in the late 1950s and early 1960s. From the robot was born to the beginning of this century 1980s, the robot technology has experienced a slow development process for a long time. By 1990s, with the rapid development of computer technology, microelectronics technology, network technology and so on, the robot technology has been developed rapidly.

Robots have three stages of development. The first generation of robots, also called the teaching and playback robot, it is through a computer, to control a number of degrees of freedom of the machine. By teaching stored procedures and information, read information at work, and then issued an order.

In the late 1970s, people started to study the second-generation robot, called robot with the feeling. The feeling of the robot is similar to the feeling of some kind of function, such as force sense, touch, slip, vision, hearing, etc.

The third generation of robot, we are pursuing the ideal of the most advanced stage, called intelligent robots. Just tell it what to do, don't tell it how to do it, it will be able to complete the movement, with the perception of thinking and human communication and other functions. With the development of science and technology, the concept of intelligence is more and more abundant, and its connotation is more and more wide.

3. Robots classification

(1) Industrial robots

The industrial robot is the most mature and widely used category of a robot. Japanese ranks first in industrial robots, becoming the kingdom of robots. The United States have developed rapidly. At present China entered the stage of industrialization, has developed a variety of industrial robots and has been mass production and used.

(2) Service robots

Robot has gradually shifted from manufacturing to non-manufacturing and service industries. Service industries include cleaning, refueling, ambulance, rescue, disaster relief, medical and so on.

4. Application of robot

Traditionally, robots are applied anywhere one of the 3Ds exists in any job which is too dirty, dangerous, and dull for a human to perform. Their function is to relieve us from danger and tedium.

Robots are particularly suitable those: human would be at significant risk, e.g. space exploration. Economically, better to use robots, where the task is menial and repetitive, and accuracy is needed, e.g. production lines. Humanitarian use where there is great risk, e.g. search and rescue in disaster zones.

Robot is mankind's right-hand man; friendly coexistence can be a reliable friend. In future, we will see and there will be a robot space inside, as a mutual aide and friend. With the development of society, the people from the heavy physical and dangerous environment liberated, so that people have a better position to work, to create a better spiritual wealth and cultural wealth.

3.3.2 Classification of Industrial Robot

1. Classification according to the application

(1) Welding robot (Fig.3-3-1)

What we say welding robot is in fact welding to produce realm to replace a welder to be engaged in the industrial robot of the welding task. Some of the welding robots are specially designed for some welding methods. Most welding robots are actually general-purpose industrial robots equipped with some welding tools.

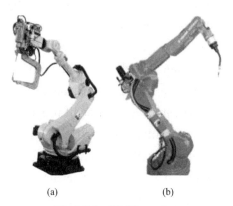

(a)   (b)

Fig.3-3-1　Welding robot

(2) Stacking robot (Fig.3-3-2)

It mainly involves the operations during storing and handling in delivery process and warehouse of the factory. Stacking robot could stack a large number of various products on the pallet according to the order in a short time.

(3) Sealing robot (Fig.3-3-3)

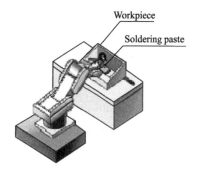

Fig.3-3-2 Stacking robot   Fig.3-3-3 Sealing robot

Install an applicator at the front end of the manipulator, for doing sealants, filler, solder coating. Sealing robot must coat the sealing parts continuously and uniformly.

(4) Cutting robot (Fig.3-3-4)

Install the cutting tools (pliers, etc.) at the front-edge of the manipulator to do the cutting.

(5) Loading and unloading robot (Fig.3-3-5)

Load raw workpieces on CNC machine and unload the finished workpieces after processing.

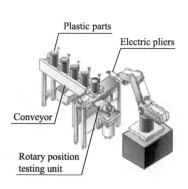
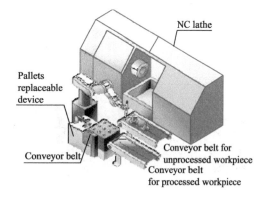

Fig.3-3-4 Cutting robot   Fig.3-3-5 Loading and unloading robot

2. Classification according to the programming coordinate system

(1) Rectangular coordinates robot (Fig.3-3-6)

Features:

① Low moving speed and easy to control;

② Operating range is less than floor space.

Application: It is suitable for loading and unloading workpieces in the assembly line.

(2) Cylindrical coordinate robot (Fig.3-3-7)

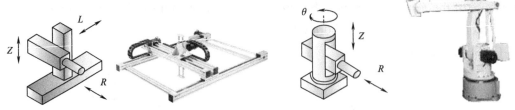

Fig.3-3-6　Rectangular coordinates robot　　　　Fig.3-3-7　Cylindrical coordinate robot

Features:

① Action range is not limited to the front;

② Some complex movements like reciprocating motion are difficult to do;

③ It has rotary function.

Application: It is suitable for mechanical workpiece installation, loading and unloading operations and other packing operations.

(3) Polar coordinates robot (Fig.3-3-8)

Features:

① The mechanical arm can rotate up and down when it operates lower or higher than the robot body;

② It can do some reciprocating motion operation;

③ The weight of the workpiece carried by the robot is less than other types of robots.

Application: It is suitable for complex space operation such as point welding, painting and curved surface simulation operation.

(4) Prosthetic robot (Fig.3-3-9)

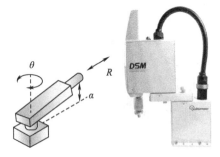
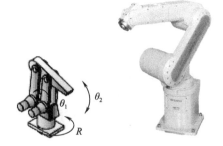

Fig.3-3-8　Polar coordinates robot　　　　Fig.3-3-9　Prosthetic robot

Features:

① It can complete complex actions;

② The action space is larger than floor space;

③ It is less accuracy, low rigidity and low weight to be carried;

④ Operation is more complex.

Application: It is suitable for assembly operations and complex curved surface follow-up operations.

3.3.3 Application of Welding Robot

The welding processing requirements of welder must have the skilled operation skills, abundant fulfillment experience, steady welding level. It is still a kind of labor condition bad, many smoke and dust, thermal radiation big, high risk of work. The emergence of the industrial robot makes people naturally thought first of using the robot instead of manual welding to reduce the welder's labor strength, and can also ensure welding quality and production efficiency at the same time.

According to incomplete statistics, there are about half of industrial robots in the world in the field of industrial robots used in various forms of welding processing. There are two main ways of welding robot application, namely, spot welding and electric arc welding.

1. Spot welding robot [Fig.3-3-1(a)]

The application of industrial robots in the field of welding is the earliest starting from the resistance spot welding of the automobile assembly line. The reason is that resistance spot welding process is relatively simple, easy to control, and does not require the seam tracking, the robot's precision and repeatability of the control requirements are relatively low. Application of welding robot in the automobile assembly line of automobile assembly welding can greatly improve the productivity and quality of welding, and has the characteristics of flexible welding, as long as the program change, can be on the same production line for different models of welding assembly.

2. Electric arc welding robot [Fig.3-3-1(b)]

In electric arc welding process, the welding workpieces due to partial melting and cooling deformation, weld trajectory will therefore change. When manual welding, the experienced welder can according to the eyes observed actual weld position to adjust the welding torch position, posture and walking speed, in order to adapt to the change of weld trajectory. Electric arc sensor is developed and applied in robot welding, which makes the welding seam tracking and control of welding seam to be solved in a certain degree.

Electric arc welding robot is the biggest characteristic of flexible, can be programmed to change the welding trajectory and welding sequence, so the most suitable for the welding variety of workpieces, short welding seam and much, more complex products. Electric arc welding robot is not only used in the automobile manufacturing industry, but also can be used in other manufacturing industries involving arc welding, such as shipbuilding, locomotives, boilers, heavy machinery, etc. Therefore, the application of arc welding robot is becoming wider and wider, and the number of robots is larger than that of spot welding robot.

New Words and Phrases

categories [ˈkætɪɡərɪz]	n. 种类，类别(category 的名词复数); 派别
directive [dəˈrektɪv]	n. 指令，命令
abundant [əˈbʌndənt]	adj. 大量的，充足的，丰富的
connotation [ˌkɒnəˈteɪʃn]	n. 内涵，含义
interdisciplinary [ˌɪntəˈdɪsəplməri]	adj. 跨学科，各学科间的
tedium [ˈtiːdɪəm]	n. 单调乏味，令人生厌，冗长
coexistence [ˌkəʊɪɡˈzɪstəns]	n. 共存，并立
warehouse [ˈweəhaʊs]	n. 仓库，货栈
reciprocating [rɪˈsɪprəkeɪtɪŋ]	adj. 往复的，交替的，摆动的
prosthetic [prɒsˈθetɪk]	n. 关节 adj. [医]义肢的，假体的
trajectory [trəˈdʒektəri]	n. [物]弹道，轨道；[几]轨线
The Robot Institute of America	美国机器人协会
microelectronics technology	微电子技术
disaster relief	赈灾，灾难援助
disaster zones	灾区
mutual aide	互相的助手
spot welding	点焊
electric arc welding	电弧焊
seam tracking	焊缝跟踪

Exercises

Ⅰ. **Match column A with column B.**

A	B
冷却变形	resistance spot welding
焊接顺序	cooling deformation
电弧焊	welding torch
焊缝跟踪	welding sequence
焊枪	electric arc welding
电阻点焊	seam tracking

Ⅱ. **Mark the following statements with T (true) or F (false).**

(   ) 1. A robot is a programmable, multi-function device.

(   ) 2. In the late 1970s, people started to study the third-generation robot.

(   ) 3. Polar coordinates robot can do some reciprocating motion operation.

(   ) 4. The application of industrial robots in the field of welding is the earliest starting from the electric arc welding.

(   ) 5. The spot welding robot can be used in the welding workpieces, the weld seam short and more complex products.

III. Answer the following questions briefly according to the text.

1. What are the three stages of robot development?
2. What is the application of the rectangular coordinates robot?
3. What's definition of the robot?
4. What is the function of the welding robot?
5. Who makes the welding seam tracking in electric arc welding process?

 Part B

Reading Material

Development Trend of Welding Robot

① Based on the application of modern design methods such as finite element analysis, modal analysis and simulation design, the optimal design of the robot's operating mechanism is realized.

② To explore new high strength lightweight materials, to further improve the load/weight ratio.

③ Using advanced reducer and AC servo motor, so that the robot has almost become a maintenance free system.

④ The mechanism is developing towards the modular and reconfigurable.

⑤ The structure of the robot is getting clever, control system smaller and smaller, twos just turn a direction development toward the integral whole.

Task 4   Automatic Production Line

 Part A

Text

3.4.1   APL Overview

1. Definition of APL

APL is the abbreviation of automatic production line. To complete a predetermined processing task, the APL comprehensively combines and applies technologies of machinery, controlling, sensing, driving, Internet and human-machine interface and integrates various machining devices in accordance with process sequence with auxiliary devices and associates all the actions by controlling hydraulic pressure, pneumatic pressure and electrical systems. CIMS (computer integrated manufacturing system) will be a perfect condition for development of the APL.

2. Development of APL

In 1920s, with the development of the automobile, rolling bearing, small motor, sewing machine, etc. the automatic production line began to appear in the machine manufacture. The first appearance is the combination of machine tool automatic production line. In 1943 the United States Ford automobile company and Krause company jointly developed an automatic production line. Metal materials or semi-finished products on the production line are automatically moved in sequence and processed continuously. In the middle of 1950s, the development of automatic production line of single variety and mass production reached its peak.

Since 1970s, in addition to the use of automatic machine and combination machine tools in automatic production line, casting, forging, welding, heat treatment, plating, painting and assembly process began to realize automation. In recent years, in the industry 4.0, driven by "made in China 2025", with the rapid development of industrial robots, automatic production lines began to know and apply by the majority of enterprises.

3. Characteristics of APL

APL is characterized by its comprehensiveness and systematicness. Comprehensiveness refers to an organized combination of various technologies in mechanics, electricity and electronics, sensors, PLC control, interfaces, driving, network communications and touch screen configuration programming, etc. and comprehensively applying them to production equipment. Systematicness refers to the production line of sensor detection, transmission and processing, control, implementation and drive institutions under the control of PLC coordinated and orderly work, organically integrated together.

3.4.2 YL-335B APL

1. The basic structure and function of YL-335B

The YL-335B equipment (Fig.3-4-1) is composed of feeding unit, transmission unit, processing unit, assembly unit and sorting unit. These units are installed in the aluminum alloy rail type training platform. Each workstation is set up a PLC to bear its control tasks, each PLC through the RS-485 serial communications to achieve interconnection, constitute a distributed control system.

Module 3 Application of Mechanotronics Device

The YL-335B APL function: The workpiece in the feeding unit storage bin is sent to the material table for processing unit, finishing process operations, machined workpiece sent to the material table for assembly unit, and then take a small cylindrical workpiece embedded in different positions within the color assembly unit to the workpiece material on the platform, finished product after assembly to refining unit and then output, refining unit according to the workpiece material, color sorting.

2. The work unit

(1) The feeding unit (Fig. 3-4-2)

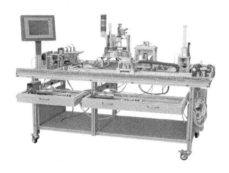

Fig. 3-4-1 The YL-335B equipment　　　　　　Fig. 3-4-2 The feeding unit

The feeding unit mainly includes workpiece storehouse, locking devices and pushing-out devices. The basic function of feeding unit is automatically push out a workpiece to be processed from the feeding bin to the material platform in order to allow manipulator of the delivery unit to pick and take the workpiece to other units.

(2) The transmission unit (Fig. 3-4-3)

The transmission unit mainly includes linear moving and workpiece picking and delivering devices. Basic function of the transmission unit is to realize the accurate positioning of the material table to a designated unit. Grasp a workpiece and take it to a determined place and drop it.

(3) The processing unit (Fig. 3-4-4)

Fig. 3-4-3 The transmission unit　　　　　　Fig. 3-4-4 The processing unit

The processing unit mainly includes workpiece moving devices and workpiece processing devices. Basic function of the processing unit is to deliver a workpiece from the material table in the unit to the pressing mechanism below for once pressing. Then deliver the workpiece back to the material table, waiting to be gripped by the manipulator.

(4) The assembly unit (Fig. 3-4-5)

The assembly unit mainly includes assemble workpiece storages and workpiece carrying devices. Basic function of the assembly unit is to embed workpiece of small columns in black or white color from feeding bin into the processed workpiece.

(5) The sorting unit (Fig. 3-4-6)

The sorting unit mainly consists of a belt conveyer and a finished product device. Basic function of the sorting unit is to sort for the processed and assembled workpiece delivered from the previous unit and allow the workpiece to be carried away through its respective chute according to its color.

(6) Control system (Fig. 3-4-7)

  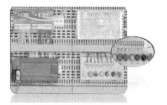

Fig. 3-4-5 The assembly unit   Fig. 3-4-6 The sorting unit   Fig. 3-4-7 Control system

YL-355B adopts PLC of Siemens S7200 series to respectively control five units as: feeding unit, transmission unit, processing unit, assembly unit, and sorting unit. PPI serial bus is adopted for communications among them.

(7) Power supply

The external power supply to YL-335B is of three-phase and five-wire AC 380V/220V.

3.4.3 The Core Technology of APL

It is common that the PLC application technology, electro technics and electronic technology, sensor technology, interface technology, network communications technology and configuration technology, etc., which are just like the human sensory system, motion system, brain and nervous system, are used in the automatic production line.

1. Transmission and power technology

In the APL, there are a lot mechanical movement controls, just like the human hands and feet, to complete the mechanical movement and action. In the APL driving devices used as power sources include motors, pneumatic devices and hydraulic devices.

(1) Mechanical transmission technology

Mechanical transmission is the transmission of power and motion by mechanical means. Mechanical transmission is widely used in mechanical engineering. Common mechanical transmission mode is belt transmission, chain transmission, gear transmission, worm gear transmission and other forms.

(2) Electric control technology

Servo motor, also known as operating motor, is used as the actuator in the automatic control system to transform the received electrical signal into the angular displacement or angular velocity in a motor shaft and then send this out. Servo motor mainly includes two kinds, DC and AC servo motor. Their main features are that, when the signal voltage is zero, there comes no rotation, and the revolution speed will fall evenly along with the torque increase. AC servo motor is a brushless motor, which can be divided into synchronous motor and asynchronous motor. The former is generally used in motion control system. Large power and inertia makes it suitable for low speed running smoothly.

(3) Pneumatic control technology

There are a number of pneumatic components, including air pump, filter pressure reducing valve, one-way electrical control valve, two-way electrical control valve, cylinder and so on.

Pneumatic system uses compressed air as working medium to transmit energy and signal. Air compressor converts mechanical energy which it is output by motor or other motor into the air pressure energy, and then with the control elements and auxiliary elements, through the implementation components the air pressure energy can be changed into mechanical energy, thus completing the linear or rotary motion.

2. Sensor and detection technology

When the workpiece goes into the sorting unit of the APL, it can be clearly observed by human eyes, However, how could the APL do it? How could we equip the APL with the function of human eyes? Just like the human sense organs such as eyes, ears and nose, the senor is the detecting component in the APL. It can feel the measured object and can convert it into the electrical signal and output it according to certain laws.

Sensors commonly used in the APL include proximity switch, displacement measuring sensor, pressure measuring sensor, flow measuring sensor, the temperature and humidity testing senor, compositions detecting sensor, image detecting sensor and many other types.

3. PLC technology

At every unit of APL, there is a PLC (Programmable Logic Controller). It is just like our brain, thinking each action, each movement, each way. It can also command the manipulator gripper according to the program. It is the core components of the APL.

4. Communications technology

In modern APL, the controlling devices in different working unit do not work independently. But by means of communications, exchange information between each other to form a whole. Therefore, the control ability and reliability of the equipment are improved, and the centralized processing and decentralized control are realized.

5. Human-machine interface technology

The human-machine interaction device provided by the human-computer interaction mode is like a window, which is the dialogue window between the operator and the PLC. The status of PLC, the current process data and fault information are graphically displayed by human-machine interface.

New Words and Phrases

impetus ['ɪmpɪtəs]	n. 动力，势头，促进
sorting ['sɔːtɪŋ]	n. 资料排架，分类
interconnection [ˌɪntəkə'nekʃn]	n. 互连
constitute ['kɒnstɪtjuːt]	vt. 构成，制定，设立
manipulator [mə'nɪpjuleɪtə(r)]	n. 操作者，操纵者，机械手
nervous ['nɜːvəs]	adj. 神经质的，紧张不安的，焦虑的
torque [tɔːk]	n. 扭转力，转（力）矩
brushless [b'rʌʃlɪs]	adj. 无刷的
synchronous ['sɪŋkrənəs]	adj. 同时存在（发生）的，同步的
inertia [ɪ'nɜːʃə]	n. 惯性，惰性
proximity [prɒk'sɪməti]	n. 亲近，接近，邻近
serial communication	串行通信
storage bin	储存箱，贮料仓，料仓
feeding unit	进给单元
pushing-out devices	推出装置
delivery unit	输送单元
centralized processing	集中处理
decentralized control	分散控制

Exercises

I. Match column A with column B.

A	B
自动化生产线	human-machine interface
人机界面	decentralized control
可编程控制器	sorting unit
分炼单元	assembly unit
装配单元	APL
分散控制	PLC

II. Mark the following statements with T (true) or F (false).

(   ) 1. PLC is the abbreviation of automatic production line.

(   ) 2. Each workstation is set up a PLC to bear its control in YL-335B.

(   ) 3. The feeding unit mainly includes workpiece moving devices and workpiece processing devices.

(   ) 4. In the APL, driving devices used as power sources include motors, pneumatic devices and hydraulic devices.

(   ) 5. Common mechanical transmission mode is air pump, filter pressure reducing valve, one-way electrical control valve, two-way electrical control valve, cylinder, and so on.

III. Answer the following questions briefly according to the text.

1. What is the main characterized of the APL?
2. What is the function of the delivery unit?
3. What are the components of the feeding unit?
4. Which two kinds are the servo motors divided into?
5. Who can realize the centralized processing and decentralized control in APL?

 Part B

Reading Material

Computer Integrated Manufacturing System (CIMS)

CIMS is the abbreviation of computer integrated manufacturing system. It is produced with the development of CAD and CAM. CIMS describes a new approach to manufacturing management, and corporate operation. Although CIMS contains many advanced manufacturing technology, such as robots, CNC, CAD, CAM, CAE and just-in-time production, but it goes beyond these techniques. CIMS is a kind of real flexible manufacturing system. It can manufacture various kinds of parts and components in batches without changing the structure of the system. CIMS includes software and automation systems needed to complete the entire process. It includes product design, system programming, production cost estimation, product actual manufacturing, instruction input, inventory tracking and actual production cost analysis.

Task 5　Mechanotronics Device Safety and Maintenance Technology

 Part A

Text

3.5.1　Safety Marks

Safety marks are usually placed on mechanotronics device in locations where hazards exist. We must understand the explanation of each safety marks. Common safety marks are shown in Table 3-5-1.

Table 3-5-1　Common safety marks

Safety marks	Safety marks
High pressure risk	Pay attention to dust
Pay attention to safe	Beware of high temperature
Be careful of noise	Warning forklift
Warning mechanical injury	Warning overhead load

(continued)

safety marks	safety marks
Caution falling objects	Warning splinter
No switching, people work	No climbing, high pressure danger
Prohibit the operation, it is working	No approaching
Must wear a safety cap	Must wear protective glasses

3.5.2 Labor Protection

Labor protection is a general term for the legislation, organization and technical measures taken by the state and the unit to protect the safety and health of workers in the process of labor production. The purpose is to create a safe labor protection for workers, health and comfortable working conditions, eliminate and prevent the possibility of labor in the production process of casualties, occupation disease and acute poisoning in order to protect workers occupation, healthy state to participate in social production. Commonly used protective measures are as follows.

1. Eyes protection

Using eyes protection in the machine shop is the most important safety rule of all. Metal chips and shavings can fly at high speed and cause serious eyes injury.

2. Hazardous noise protection

Noise hazards are very common in the machine shop. High intensity noise can cause permanent loss of hearing. Although noise hazards cannot always be eliminated, hearing loss is avoidable with earmuffs, ear plugs, or both.

3. Foot protection

The floor in a machine shop is often covered with razor-sharp metal chips, and heavy stock may be dropped the feet. Therefore, safety shoes or a solid leather shoe must be worn at all times. These have a steel plate located over the toe and are designed to resist impact. Some safety shoes also have an instep guard.

Electrical work must wear qualified insulation shoes. Insulation shoes as engaged in electrical work safety auxiliary tool, can effectively reduce the chance of electric shock.

3.5.3 Daily Maintenance

Daily maintenance is one of the most important maintenance ways. If you often maintain your device and read the maintenance manual, you will deal with many minor issues in your work. These required specifications must be followed in order to keep your device in good working order and protect your warranty.

1. Checking the external view

① Top off coolant level every eight hour shift.
② Check lubrication tank level of slide way.
③ Clean chips from slide way covers and bottom pan.
④ Clean chips from tool changer.
⑤ Check oil level in gearbox. Add oil until oil begins dripping from over flow tube at bottom of sump.
⑥ Inspect slide way covers for proper operation and lubricate with light oil, if necessary.
⑦ Check all hoses and lubrication lines for cracking.
⑧ Check oil filter and clean out residue at bottom of filter.

2. Checking the inside of the control unit

Check that the follow troubles have been eliminated:
① Cable connectors are loosened.
② Cable is damaged.
③ Installing screws are loosened.
④ Screws used to attach amplifier are loosened.
⑤ The cooling fan operates abnormally.
⑥ Printed circuit boards have been inserted abnormally.

3.5.4 Fault Diagnosis

1. Fault diagnosis technology

The task of fault diagnosis consists in determining the type, size and location of the most possible faults, as well as the time of detection.

For further improvement of the reliability and safety of device, the automatic early detection and localization of faults are very important. The conventional approach is to monitor some important variables like temperature, pressure, vibration and to generate alarms if certain limits are exceeded.

The operation of technical processes requires increasingly advanced supervision and fault diagnosis to improve reliability, safety and economy.

2. Corrective action for failures

When a failure occurs, it is important to correctly grasp what kind of failure occurred and take appropriate action to promptly repair the device.

Check for the failures according to the following procedure (Fig.3-5-1).

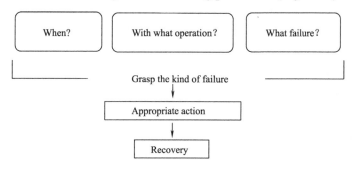

Fig. 3-5-1　Corrective action for failures

(1) When did the failure occur
① Date and time.
② Occurred during operation?
③ Occurred when the power was turned on?
④ Were there any lightening surges, power failure or other disturbances to the power supply? How many times has it occurred?
⑤ Only once?
⑥ Occurred many times? (How many times per hour, per day or per month?)

(2) With what operation did it occur
① What was the systems mode when the failure occurred?
② Does the same operation cause the same failure? (Check the repeatability of the failure.)
③ Occurred during data input/output?
④ For a failure related to feed axis servo?
a. Occurred at both low feed rate and high feed rate?
b. Occurred only for a certain axis?
⑤ For a failure related to spindle.

When did the failure occur? (during power-on, acceleration, deceleration or constant rotation)

(3) What failure occurred
① Which alarm was displayed on the alarm display screen?
② Is the display screen correct?
③ Is the machining dimensions incorrect?
a. How large is the error?
b. Is the position display screen on the CRT correct?
c. Are the offsets correct?
(4) Other information
① Is there noise origin around device?
② Is it taken any countermeasures for noise in device side?
③ Check the following for the input power supply voltage:
a. Is there variation in the voltage?
b. Are the voltages different depending on the phase?
c. Is the standard voltage supplied?
④ How high is the ambient temperature around the control unit?
⑤ Has excessive vibration been applied to the control unit?

New Words and Phrases

forklift ['fɔːklɪft]	n. 叉车
splinter ['splɪntə(r)]	n. 碎片，尖片
casualties ['kæʒjuəlti]	n. 伤亡者(casualty 的名词复数)，牺牲品，受害者
razor-sharp	adj. 锋利的，犀利的
instep ['ɪnstep]	n. 脚背，脚背形的东西
electric shock	n. 电击，触电
maintenance ['meɪntənəns]	n. 维护，维修，维持，保持，保养，保管
fault [fɔːlt]	n. 缺点，缺陷，过错，责任；[电]故障
diagnosis [ˌdaɪəɡ'nəʊsɪs]	n. 诊断，诊断结论，判断，结论
earmuffs ['ɪəmʌfs]	n. 防噪声耳罩
safety mark	安全标识
general term	泛称，总称
occupation disease	职业病
acute poisoning	急性中毒
ear plug	防噪声耳塞
instep guard	护脚背
insulation shoes	绝缘鞋
daily maintenance	日常保养与维护
fault diagnosis	故障诊断

lightening surges　　　　　　　　打雷

Exercises

Ⅰ. Match column A with column B.

A	B
故障诊断	safety mark
职业病	daily maintenance
安全标识	occupation disease
日常保养	fault diagnosis
劳动保护	acute poisoning
急性中毒	labor protection

Ⅱ. Mark the following statements with **T** (true) or **F** (false).

(　) 1. Safety decals are usually placed on mechanotronics devices in locations where hazards exist.

(　) 2. Metal chips can fly at slow speed and can't cause serious eye injury.

(　) 3. Some safety shoes also have function for instep guarding.

(　) 4. Screws used to attach amplifier may be loosened.

(　) 5. It is important to correctly grasp what kind of failure occurred and take appropriate action to promptly repair the device when a failure occurs.

Ⅲ. Answer the following questions briefly according to the text.

1. Where is the general safety mark posted?
2. What is the role of labor protection?
3. What are the common protective measures?
4. What can cause permanent loss of hearing in a workshop?
5. What fault diagnosis includes?

 Part B

Reading Material

Color Markings

All maintenance shops and work areas should be marked with the correct colors to identify hazards, exits, safe walkways and first-aid station. It is acceptable to use material other than paint, such as decals and tapes, in the appropriate, similar colors. Listed below are the main colors authorized for use in maintenance shops.

Red color markings should be used to identify the following equipment or locations:

(1) Fire alarm boxes
(2) Fire blanket boxes
(3) Fire pumps
(4) Emergency stop buttons for electrical machinery
(5) Emergency stop bars on hazardous machines
(6) Caution signs

Green color markings normally on a white color background should be used for the following equipment or locations:

(1) First-aid equipment
(2) First-aid dispensaries
(3) Safety starting buttons on machinery

Orange markings are used to designate dangerous components of machines or energized equipment.

课文翻译

模块3 机电一体化设备应用

任务1 数控机床

 Part A

3.1.1 数控机床概述

1. 数控机床的工作过程

首先要把加工零件需要的工艺信息以程序的形式记录下来,存储在某些载体上,输入到数控装置中,由数控装置处理程序,发出控制信号指挥伺服系统驱动机床、协调机床的动作,使其产生主运动和进给运动的一系列机床运动,完成零件的加工。

2. 数控机床的优点

数控机床和传统机床相比具有很多优点,它们是:
① 同一机床,一次装夹可完成多种操作。
② 由于可以实现多轴联动,切削不规则形状时,不必使用成形刀具。
③ 由于数控机床的精度高,受操作者影响小,废品率明显降低。
④ 生产效率显著提高。

3. 数控机床的缺点

数控机床也有缺点,它们是:
① 价格昂贵。
② 必须由高级技术人员来编程、安装、操作和维护。

3.1.2 数控车床

1. 数控车床概述

数控车床是计算机数字控制车床。数控车床上最主要的操作有:车端面、圆柱面车削、车退刀槽、切断、钻孔、车孔、螺纹加工等。

2. 数控车床的组成部件（图3-1-1）

床身固定在底座上,其上安装着车床的各主要部件,并使它们在工作时保持准确的相对位置。

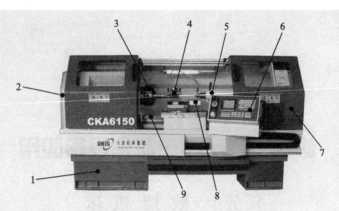

图 3-1-1　数控车床组成部件

1—床身；2—主轴箱；3—三爪卡盘；4—转塔刀架；5—尾座；6—操作面板；7—防护罩；8—刀架电动机；9—导轨

主轴箱固定于床身的最左边，主要用于支承并带动主轴，以实现机床的主运动。

刀架滑板由纵向（Z 向）滑板和横向（X 向）滑板组成。纵向滑板安装在床身导轨上，沿床身实现纵向（Z 向）运动；横向滑板安装在纵向滑板上，沿纵向滑板上的导轨实现横向（X 向）运动。刀架滑板可使安装在其上的刀具实现纵向和横向进给运动。

转塔刀架安装在机床的刀架滑板上，用于装夹各种刀具。加工时可根据加工要求自动换刀。

尾座安装在床身导轨上，可沿导轨进行纵向移动。尾座主要用于安装顶尖，在加工中对工件进行辅助支承。尾座上也可安装钻头、铰刀等刀具进行孔加工。

防护罩安装在机床底座上，加工时保护操作者的安全和保护环境的清洁。

机床的液压传动系统用来实现机床上的一些辅助运动，主要是实现机床主轴的变速、尾座套筒的移动及工件自动夹紧机构的动作。

机床控制系统主要由数控系统（包括数控装置、伺服系统及可编程控制器）和机床的电气控制系统组成，它能够完成对机床的自动控制。

3. 数控车床的分类

（1）按照主轴位置分类

按照主轴位置，数控车床可分为卧式数控车床（图 3-1-1）和立式数控车床（图 3-1-2，略）。立式数控车床用于回转直径较大的盘类零件车削加工。卧式数控车床用于加工各种轴类、套筒类和盘类零件的回转表面，如内外圆柱面、圆锥面、螺纹面。

（2）按照控制系统特征分类

按照控制系统特征，数控车床可分为经济型数控车床、全功能型数控车床、车削中心和 FMC（柔性制造单元）车床。

经济型数控车床（图 3-1-3，略）一般是在普通车床的基础上进行改进设计，装备开环伺服系统，其控制部分采用单片机。此类车床结构简单，价格低廉，但与其他数控车床比较没有刀尖圆弧半径自动补偿和恒线速度切削等功能。

全功能型数控车床，通常简称数控车床，又称标准型数控车床（图3-1-4，略）。其系统是标准型的，带有高分辨率的液晶显示器，具有图形仿真、刀具补偿、通信或网络接口、多轴联动等标准功能。全功能型数控车床采用闭环或半闭环控制的伺服系统，具有高刚度、高精度和高效率等特点。

车削中心（图 3-1-5，略）以全功能型数控车床为主体，配有刀库、换刀装置、分度装置、铣削装置和机械手等部件，以实现多工序复合加工。工件一次装夹后，在车削中心可完成回转类零件的车、铣、钻、铰、攻螺纹等多种加工工序。其效率和自动化程度比全功能型数控车床高，但价格较贵。

FMC 车床（图 3-1-6，略）是由数控车床、机器人等构成的柔性加工单元。它能够实现工件搬运、装卸的自动化和加工调整准备的自动化。

3.1.3 数控加工中心

1. 数控加工中心概述

数控加工中心是具有自动换刀能力和旋转刀库的多功能数控机床。加工中心的主要优势是提高了生产率和扩大了加工范围。在一台加工中心上可以实现钻孔、车削、铰孔、镗孔、铣削、轮廓加工和攻螺纹等多种操作。大多数工件通常能在一台加工中心上一次装夹完成全部加工。

2. 数控加工中心的组成部件

数控加工中心主要部件是床身、滑座、立柱、工作台、伺服系统、主轴、换刀装置和机床控制单元（图 3-1-7，略）。

床身通常由优质铸铁制成，铸铁为刚性机床提供了执行重载加工任务和保持高精度加工的能力，床身用来支撑所有部件。

工作台安装在床身上，为加工中心提供沿 X 轴的线性移动。

滑座安装在经过硬化和研磨处理的床身导轨上，为加工中心提供沿 Y 轴的线性移动。

立柱安装在滑座上，立柱为加工中心提供沿 Z 轴的线性移动。

伺服系统包括伺服电动机、滚珠丝杠和位置反馈编码器，它能使 X 轴、Y 轴或 Z 轴快速而准确地进行定位。

换刀装置能够存储一定数量的预置刀具，零件加工程序可以自动调用这些刀具。换刀时间通常只有 3～5 s。

机床控制单元（MCU）是用于存储和处理输入程序的计算机。MCU 因制造商的设计参数不同而不同，新型的 MCU 变得更加复杂，使得机床可靠性更高，整个加工操作更少地依赖于人的技巧。

3. 加工中心分类

加工中心可以分为立式加工中心（图 3-1-8，略）、卧式加工中心（图 3-1-9，略）、万能加工中心（图 3-1-10，略）和龙门加工中心（图 3-1-11，略）。立式加工中心应用广泛，主要用于加工平板零件。卧式加工中心应用也十分广泛，特别是加工大型的、箱体的和重型的零件。万能加工中心同时装备有垂直和水平主轴，具有多种功能，能够同时加工件的所有表面。龙门加工中心适用于加工大型或形状复杂的精密工件，如汽轮机叶片等。

4. 刀库类型

（1）直线式刀库（图 3-1-12，略）

直线式刀库中刀具呈直线排列，结构简单，存放刀具数量一般为 6~8 把，多用于数控车床，也可用于数控钻床。

（2）圆盘式刀库（图 3-1-13，略）

存刀量为 6~60 把，并且有多种形式。刀具布置形式有径向布置、轴向布置和伞状布置。

（3）链式刀库（图 3-1-14，略）

链式刀库也是应用广泛，其存刀量一般为 30~120 把。

（4）箱式刀库（图 3-1-15，略）

箱式刀库容量较大，刀库的整体更换较为便捷，刀库占地面积小。但选刀和取刀动作复杂，因此，很少用于单机加工中心，多用于柔性制造系统的集中供刀系统。

3.1.4 数控机床安全操作规程

金属切削操作中安全性一直受到关注。由于计算机数控设备自动化程度高并且速度快，所以它是一个危险源。为了防止人员伤害和对设备的损坏，必须找出存在危险的根源，且操作人员必须提高警惕。主要潜在的危险包括：旋转部件，如主轴、主轴装卡的刀具、卡盘、卡盘装卡的工件、装卡着刀具的刀塔及旋转的夹具装置；运动部件，如加工中心的工作台、车床拖板、尾架顶尖和刀塔；程序错误，如 G00 指令的不正确使用而引起坐标值错误，产生意想不到的快速移动；设置或改变偏移值时出错，可能导致刀具与工件或刀具与机床之间的碰撞；随意地更改已验证的程序，也会引起机床产生危险动作。为了减少或避免危险，应尽量遵循以下保护措施：

① 使用机床制造商提供的机器原有防护罩。

② 带上安全眼镜、手套，穿上合适的衣服和鞋。

③ 不熟悉机床操作控制前不要开动机床。

④ 运行程序之前，确认零件已被正确夹紧。

⑤ 验证一个程序时，应遵循下列安全步骤：

a. 启用机床锁定功能运行程序，检查程序中的语法错误和几何轨迹。

b. 使用"快速倍率"开关降低速度或空运行程序。

c. 采用单程序段执行来确定程序中的每一行。

d. 刀具切削时，用"进给倍率"开关来减慢进给速率，防止超负荷切削。

⑥ 禁止用手处理切屑及用切屑钩子弄断长而卷曲的切屑。如果需要彻底清除切屑，应当关闭机床。

⑦ 尽可能保持刀具伸出短些，因为它可能成为导致刀片折断的振动源。

⑧ 更换刀具、替换刀片或清理切屑时一定要关闭机床。

⑨ 在未得到主管许可的情况下不得擅自更改程序。

⑩ 如果你有任何与安全有关的担忧，立即通知你的技术指导或主管。

Part B

数控机床的分类

按照刀具的运动方式,数控机床分为点位控制和轮廓控制(或连续路径控制)。点位控制系统将刀具从一点移动到另一点。这种控制系统最简单的例子是钻床,在钻头从一点移动到下一点的过程中,路径和进给无关紧要,不控制从起点到终点的运动轨迹。轮廓控制系统能够在至少两个坐标轴上同时调整工作台(或主轴)的进给,这种控制方式需要复杂的控制和驱动系统。

按照机床控制环,数控机床分为开环和闭环控制。开环控制系统不给控制单元提供位置反馈信息。开环控制系统的优点是成本低,缺点是很难检测位置误差。闭环控制系统精度高,通常采用交流、直流或液压伺服电动机。

任务 2 数控电加工机床

Part A

3.2.1 数控电火花成形机床

1. 数控电火花成形加工的原理(图 3-2-1)

放电加工又称电火花加工,是基于两电极之间产生的电火花的腐蚀作用实现的。在工具电极与工件之间连接脉冲电源,利用自动进给调节装置使工具电极与工件始终保持一个很小的放电间隙。在脉冲电流的作用下,产生火花放电。工件与电极表面上的凸峰处电流密度很大,产生局部高温,因此凸峰将首先被熔化、汽化,形成微小凹坑,这又会形成新的凸峰;下次脉冲放电时,又会在新的凸峰处蚀除金属,熔化的金属以粉末状散布于工作液中被带走,过滤掉。这样不断地脉冲放电,就可将工具电极的形状复制在工件上,实现成形加工。

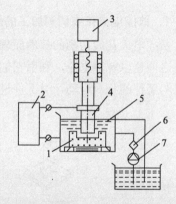

图 3-2-1 数控电火花成形加工的原理
1—工件;2—脉冲电源;3—伺服机构;4—电极;
5—电解液;6—电解液过滤器;7—泵

2. 数控电火花成形加工机床的组成

数控电火花成形加工机床(图 3-2-2,略)是一种高精度的自动化加工机床。它由机床本体、脉冲电源、自动进给调节装置、数控系统、工作液循环系统等组成。机床主轴上装有工具电极(正极),工件(负极)夹装在工作台上。在数控系统控制下,Z 向伺服电动机通过滚珠丝杠带动主轴上下运动,使工具电极和工件之间保持稳定的放电间隙,实现电蚀加工过程。工作台由伺服电动机通过滚珠丝杠实现 X、Y 向进给运动,完成指定的轨迹加工。

3. 数控电火花成形加工的特点及应用

（1）数控电火花成形加工的特点

① 数控电火花成形加工的优点：

a. 可以加工用其他传统切削刀具无法加工的复杂形状。

b. 越硬的材料加工精度越高。

c. 用传统的方法加工由于过大的刀具切削力可能损伤非常小的零件。

d. 由于加工中工具电极和工件不直接接触，因此可以加工低刚度工件以及进行细微加工，不产生变形。

② 数控电火花成形加工的缺点：

a. 材料去除速度缓慢。

b. 需要花费额外的时间和费用为电火花成形机床制作电极。

c. 由于电极的损耗，在工件上复制尖角是非常困难的。

d. 电能消耗非常高。

（2）数控电火花成形加工的应用

由于数控电火花成形加工能够解决复杂形状零件的加工问题，所以其应用领域日益扩大。数控电火花成形加工不仅广泛应用于模具制造领域，而且广泛应用于航空、小孔和微孔生产。

3.2.2 数控电火花线切割机床

1. 数控电火花线切割加工的基本原理（图3-2-3）

数控电火花线切割的基本原理是利用连续移动的细金属丝作为工作电极，并在金属丝与工件之间通以脉冲电流，利用它们之间的脉冲火花放电效应，使金属熔化或汽化，并通过电极丝与工件的相对运动，对工件进行切割成形。数控电火花线切割又称线切割。

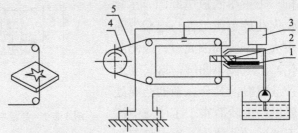

图 3-2-3 数控电火花线切割的基本原理

1—绝缘底板；2—工件；3—脉冲电源；4—丝筒；5—电极丝

2. 数控电火花线切割机床的分类

根据电极丝的运行速度和运转方式，数控线切割机床可以分为以下两类：

（1）高速走丝线数控切割机床（WEDM-HS，图3-2-4，略）

这类机床的电极丝做高速往复运动，又称往复走丝或快走丝线切割机。

（2）低速走丝线数控切割机床（WEDM-LS，图3-2-5，略）

这类机床的电极丝做低速单向运动，又称单向走丝或慢走丝切割机。

3. 数控电火花线切割机床的组成

数控电火花线切割机床主要由机床本体、脉冲电源、工作液、数控系统和机床附件等组成（图 3-2-6）。

图 3-2-6 数控电火花线切割机床的组成
1—床身；2—工作液；3—坐标工作台；4—丝架；5—数控系统；6—脉冲电源

（1）机床本体

① 床身。床身是坐标工作台、储丝机构及丝架的支撑，应具有足够的强度和刚度，床身内部可以安置电源及工作液箱。

② 坐标工作台。坐标工作台是用来安置工件，并根据控制要求，相对于电极丝做预定的相对运动的，它包括拖板、导轨及驱动装置等。

③ 丝架。丝架的作用是通过丝架上的两个导轮对电极丝进行支撑和导向，并且能使电极丝的工作部分与工作台保持一定的角度，以便实现锥度切割。双坐标联动丝架，是在丝架上增加了 U、V 两个驱动电动机，通过程序控制来实现锥度切割。

（2）脉冲电源

脉冲电源的作用是把工频交流电转换成一定频率的单向脉冲电流，以供给工件和电极丝放电间隙所需要的电能来蚀除金属，它的性能直接影响加工速度、表面质量、加工精度及电极丝的损耗等。

（3）工作液

工作液在加工时起绝缘、洗涤、排屑、冷却的作用，对切割速度、表面粗糙度、加工精度均有影响。因此，对工作液性能有以下要求：

① 具有一定的绝缘性。
② 具有较好的磁导率。
③ 冷却性能好。
④ 对环境无污染，对人体无危害。
⑤ 价格低，稳定性好。

（4）数控系统

数控系统的主要作用是在电火花线切割加工的过程中，自动控制电极丝与工件的相对运动轨迹和进给速度，实现自动加工。其主要功能是轨迹控制和加工控制。

4. 数控电火花线切割机床的特点及应用

（1）数控电火花线切割的特点

① 数控电火花线切割的优点：

a. 可以切割任何硬度的材料。

b. 可以获得很高的精度和表面质量。

c. 不存在切割应力。

② 数控电火花线切割的缺点：

a. 只能加工导电材料。

b. 加工速度慢，特别是加工高表面质量和高精度要求的零件。

（2）数控电火花线切割加工的应用

数控电火花线切割加工是电火花加工中最常见的类型之一，是一种高精度的加工方法，可以加工硬度高和用传统机械加工方法无法加工的材料。在工业上，数控电火花线切割技术应用于电子、工具和模具、航空航天、机器人和医疗等加工行业。

 Part B

电规准的选择

使用者必须根据工作要求选取电规准，包括电流、脉宽、脉冲间隔等参数。

1. 粗加工

为了获得较快的加工速度，应选择大脉冲宽度和大电流。电流选择时应考虑电极尺寸，以免单位面积电流太大；脉冲间隔从加工速度角度考虑选择应尽量小；但小脉冲间隔易造成加工条件恶化，间接造成电极损耗增大。为了获得较小的电极损耗，应选择负极性加工，即工件接负极，工具电极接正极。

2. 精加工

粗加工的主要目的是获得良好的表面粗糙度和尺寸精度。脉冲宽度要小，电流也要小；由于排屑条件恶劣，脉冲间隔应选大一些，抬刀要频繁，但高度要低。脉冲宽度应选择 80 μs 以下，以保证放电稳定。

任务 3　机　器　人

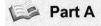 **Part A**

3.3.1　机器人概述

机器人技术涉及力学、机械学、电气液压技术、自动控制技术、传感器技术和计算机科学等科学领域,是一门跨学科的综合技术。当前,应用于工业领域的有三菱、库卡、ABB 等多个公司的机器人。

1. 机器人的定义

美国机器人协会对机器人的定义如下:机器人是一种通过不同的程序动作来完成多种工作的可编程、多功能的装置,其设计目的是用来搬运材料、零部件、工具或特殊装置。

2. 机器人的发展

第一个机器人是在 20 世纪 50 年代后期和 60 年代早期由 Devol 和 Engelberger 发明的通用机械手。从机器人诞生到 20 世纪 80 年代初,机器人技术经历了一个长期缓慢的发展过程。到了 20 世纪 90 年代,随着计算机技术、微电子技术、网络技术等的快速发展,机器人技术也得到了飞速发展。

机器人有三个发展阶段。第一代机器人,又称示教再现型机器人,它是通过一台计算机,来控制一个多自由度的机械。通过示教存储程序和信息,工作时把信息读取出来,然后发出指令。

在 20 世纪 70 年代后期,人们开始研究第二代机器人,称为带感觉的机器人。这种带感觉的机器人是类似人在某种功能的感觉,比如力觉、触觉、滑觉、视觉、听觉等。

第三代机器人称为智能机器人,是人们所追求的、理想的、最高级的阶段。只要告诉它做什么,不用告诉它怎么去做,它就能完成动作,具有感知思维和人机通信等功能。随着科学技术的发展,智能的概念越来越丰富,它内涵越来越宽。

3. 机器人的分类

(1) 工业机器人

工业机器人是最成熟,应用最广泛的一类机器人。日本在工业机器人排名第一,成为机器人的王国;美国发展也很迅速;目前中国进入产业化的阶段,已经研制出多种工业机器人,并且开始批量生产和使用。

(2) 服务机器人

机器人已经从制造业逐渐转向了非制造业和服务行业。服务行业包括清洁、加油、救护、抢险、救灾、医疗等。

4. 机器人的应用

从传统意义上来说,机器人可以用于任何存在 3D 的场合。所谓的 3D 就是在任何工作当中,对人们来说太脏(dirty)、太危险(dangerous)、太沉闷(dull)的工作。而他们的职能就是将人们从危险和沉闷中解放出来。

机器人特别适合于那些对人类极度冒险的场合,如太空探险;从经济上来说,更应使用机器人的场合——枯燥而重复的工作且要求高精度的地方,如生产流水线;从人道主义考虑,对人类有很大危害的地方,如在灾害地区进行搜索和救援。

机器人是人类的得力助手,能友好相处的可靠朋友。将来我们会看到人和机器人存在一个空间里面,成为一个互相的助手和朋友。随着社会的发展,机器人把人们从繁重的体力和危险的环境中解放出来,使人们有更好的岗位去工作,去创造更好的精神财富和文化财富。

3.3.2 工业机器人分类

1. 按照应用分类

（1）焊接机器人（图3-3-1，略）

焊接机器人是在焊接生产领域代替焊工从事焊接任务的工业机器人。焊接机器人中有的是为某种焊接方式专门设计的，而大多数的焊接机器人其实就是通用的工业机器人装上某种焊接工具而构成的。

（2）码堆机器人（图3-3-2，略）

主要在产品出厂工序和仓库的储存保管时进行作业。使用码堆机器人，就能够在短时间内按照订单将各类产品大量、迅速地堆积在托板上交付。

（3）密封机器人（图3-3-3）

在机械手前端安装涂敷头，进行密封剂、填料、焊料涂敷等作业。密封机器人必须对密封部位进行连续、均匀涂敷。

（4）切割机器人（图3-3-4）

在机械手前端装上切割工具（剪钳等）进行作业。

（5）装、卸机器人（图3-3-5）

用于在数控机床的工件夹头上安装未加工的工件，并且将加工结束后的工件取下。

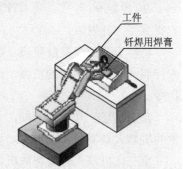

图3-3-3 密封机器人

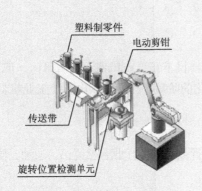

图3-3-4 切割机器人

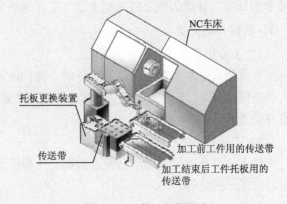

图3-3-5 装、卸机器人

2. 按照编程坐标系分类

（1）直角坐标机器人（图3-3-6，略）

特点：

① 移动速度慢，便于控制；

② 作业范围小于占地面积。

应用：适用于装配线中装卸工件。

（2）圆柱坐标机器人（图3-3-7，略）

特点：

① 作业范围不再局限于正面；

② 迂回等复杂动作难以执行；
③ 具有回转功能。

应用：适用于机械上的工件安装、装卸作业及其他包装作业。

（3）极坐标机器人（图 3-3-8，略）

特点：
① 在低于或高于机器人躯体的位置处进行作业时，机械臂可上下回转；
② 可进行某种程度的迂回作业；
③ 可搬运的工件重量小于其他形态的机器人。

应用：用于点焊、喷涂等空间位置较复杂的作业及曲面仿型加工作业。

（4）关节机器人（图 3-3-9，略）

特点：
① 可完成复杂动作；
② 作业范围大于占地面积；
③ 精度、刚性、可搬运重量较低；
④ 操作比较复杂。

应用：适用于组装作业和复杂的曲面随动等作业。

3.3.3 焊接机器人应用

焊接加工要求焊工有熟练的操作技能、丰富的实践经验、稳定的焊接水平。它还是一种劳动条件差、烟尘多、热辐射大、危险性高的工作。工业机器人的出现使人们自然而然首先想到用它代替人的手工焊接，减轻焊工的劳动强度，同时也可以保证焊接质量和提高焊接效率。

据不完全统计，全世界在役的工业机器人中大约有一半的工业机器人用于各种形式的焊接加工领域。焊接机器人应用中最普遍的主要有两种方式，即点焊和电弧焊。

1. 点焊机器人 ［图 3-3-1（a），略］

工业机器人在焊接领域的应用最早是从汽车装配生产线上的电阻点焊开始的。原因在于，电阻点焊的过程相对比较简单，控制方便，且不需要焊缝轨迹跟踪，对机器人的精度和重复精度的控制要求比较低。点焊机器人在汽车装配生产线上的大量应用大大提高了汽车装配焊接的生产率和焊接质量，同时又具有柔性焊接的特点，即只要改变程序，就可在同一条生产线上对不同的车型进行装配焊接。

2. 电弧焊机器人 ［图 3-3-1（b），略］

电弧焊过程中，被焊工件由于局部加热熔化和冷却产生变形，焊缝的轨迹会因此而发生变化。手工焊时，有经验的焊工可以根据眼睛所观察到的实际焊缝位置适时地调整焊枪的位置、姿态和行走的速度，以适应焊缝轨迹的变化。电弧传感器的开发并在机器人焊接中得到应用，使机器人电弧焊的焊缝轨迹跟踪和控制问题在一定程度上得到解决。

电弧焊机器人的最大的特点是柔性，即可通过编程随时改变焊接轨迹和焊接顺序，因此最适用于被焊工件品种变化大、焊缝短而多、形状复杂的产品。电弧焊机器人不仅用于汽车制造业，更可以用于涉及电弧焊的其他制造业，如造船、机车车辆、锅炉、重型机械等。因此，机器人电弧焊的应用范围日趋广泛，在数量上大有超过点焊机器人之势。

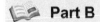 **Part B**

焊接机器人的发展趋势

① 通过有限元分析、模态分析及仿真设计等现代设计方法的运用，实现机器人操作机构的优化设计。
② 探索新的高强度轻质材料，进一步提高负载/自重比。
③ 采用先进的减速器及交流伺服电动机，使机器人操作几乎成为免维护系统。
④ 机构向着模块化、可重构方向发展。
⑤ 机器人的结构更加灵巧，控制系统愈来愈小，二者正朝着一体化方向发展。

任务4　自动化生产线

 **Part A**

3.4.1　自动化生产线概述

1. 自动化生产线的定义

APL 是自动化生产线的缩略语。自动化生产线综合应用机械技术、控制技术、传感技术、驱动技术、网络技术、人机接口技术等，通过一些辅助装置按工艺顺序将各种机械加工装置连成一体，并控制液压、气压和电气系统将各个部分动作联系起来，完成预定的生产加工任务。CIMS（计算机集成制造系统）将是自动化生产线发展的一个理想状态。

2. 自动化生产线的发展

20 世纪 20 年代，随着汽车、滚动轴承、小型电动机和缝纫机等工业的发展，机械制造中开始出现自动化生产线。最早出现的是组合机床自动化生产线。1943 年美国福特汽车公司与克罗斯公司共同研制出一条自动生产线。在生产线上的金属材料或半成品自动按顺序地移动，并连续不断地被加工。20 世纪 50 年代中期，这种单一品种、大批量生产的自动化生产线的发展达到了巅峰。

20 世纪 70 年代以来，除了用自动机床和组合机床构成自动化生产线外，铸造、锻压、焊接、热处理、电镀、喷漆和装配等过程也开始实现自动化。近年来，在工业 4.0、中国制造 2025 等的推动下，随着工业机器人的快速发展，自动化生产线开始被广大企业熟知并应用。

3. 自动化生产线的特点

自动化生产线最大特点是综合性和系统性。综合性是指机械技术、电工电子技术、传感器技术、PLC 控制技术、接口技术、驱动技术、网络通信技术、触摸屏组态编程等多种技术有机地结合，并综合应用到生产设备中。系统性是指生产线的传感检测、传输与处理、控制、执行与驱动等机构在 PLC 的控制下协调有序地工作，有机地融合在一起。

3.4.2　YL-335B 型自动化生产线

1. YL-335B 的基本结构及功能

亚龙 YL-335B 装备（图 3-4-1，略）由供料单元、输送单元、加工单元、装配单元和分拣单元组成，这些单元安装在铝合金导轨式实训台上。各工作站均设置一台 PLC 承担其控制任务，各 PLC 之间通过 RS-485 串行通信实现互连，构成分布式的控制系统。

YL-335B 型自动化生产线的功能：将供料单元料仓内的工件送往加工单元的物料台，完成加工操作后，把加工好的工件送往装配单元的物料台，然后把装配单元仓内的不同颜色的小圆柱工件嵌入到物料台上的工件中，完成装配后的成品送往分拣单元分拣输出，分拣单元根据工件的材料、颜色进行分拣。

2. YL-335B 的工作单元

（1）供料单元（图 3-4-2，略）

供料单元主要由工件库、工件锁紧装置和工件推出装置组成。供料单元的基本功能是按照需要将放置在仓料中待加工的工件自动送出到物料台上，以便输送单元的抓取机械手装置将工件抓取送往其他工作单元。

（2）输送单元（图 3-4-3，略）

输送单元主要包括直线移动装置和工件取送装置。输送单元的基本功能是实现到指定单元的物料台精确定位，并在该物料台上抓取工件，把抓取到的工件输送到指定地点然后放下。

（3）加工单元（图 3-4-4，略）

加工单元主要包括工件搬运装置和工件加工装置。加工单元的基本功能是把该单元物料台上的工件送到冲压机构下面，完成一次冲压加工动作，然后再送回到物料台上，待抓取机械手装置取出。

（4）装配单元（图 3-4-5，略）

装配单元主要包括装配工件库和装配工件搬运装置。装配单元的基本功能是完成将该单元料仓内的黑色或白色小圆柱工件嵌入到已加工的工件中。

（5）分拣单元（图 3-4-6，略）

成品分拣单元主要包括传送带输送线和成品分拣装置。分拣单元的基本功能是将上一单元送来的已加工、装配的工件进行分拣，使不同颜色的工件从不同的料槽分流。

（6）控制系统（图 3-4-7，略）

YL-335B 采用西门子 S7200 系列 PLC，分别控制供料、输送、加工、装配、分拣 5 个单元。5 个单元之间采用 PPI 串行总线进行通信。

（7）供电电源

YL-335B 外部供电电源为三相五线制 AC 380 V/220 V。

3.4.3 自动化生产线的核心技术

自动化生产线中通常用到 PLC 应用技术、电工电子技术、传感器技术、接口技术、网络通信技术、组态技术等，就像人的感官系统、运动系统、大脑及神经系统。

1. 传动与动力技术

在自动化生产线中，有许多机械运动，就像人的手和足一样，用来完成机械运动和动作。自动化生产线中作为动力源的传动装置有电动机、气动装置和液压装置。

（1）机械传动技术

机械传动是利用机械方式传递动力和运动的传动。机械传动在机械工程中应用非常广泛，常用的机械传动方式有带传动、链传动、齿轮传动、蜗轮蜗杆传动及其他传动方式。

（2）电动技术

伺服电动机又称执行电动机，在自动控制系统中，用作执行元件，把所收到的电信号转换成电动机轴上的角位移或角速度输出。伺服电动机分为直流伺服电动机和交流伺服电动机两大类，其主要特点是，当信号电压为零时无自转现象，转速随着转矩的增加而匀速下降。交流伺服电动机是无刷电动机，分为同步电动机和异步电动机。目前运动控制中一般都用同步电动机，它的功率范围大，惯量大，因而适于低速平稳运行的应用。

（3）气动技术

气动元件包括气泵、过滤减压阀、单向电控阀、双向电控阀、气缸等。

气动系统是以压缩空气为工作介质来进行能量与信号传递的。利用空气压缩机将电动机或其他原动机输出的机械能转换为空气的压力能，然后在控制元件的控制和辅助元件的配合下，通过执行元件把空气的压力能转变为机械能，从而完成直线或回转运动。

2. 传感器及检测技术

当工件进入自动化生产线中的分拣单元，人的眼睛可以清楚地观察到，但自动化生产线是如何来判别的呢？如何使自动化生产线具有人眼功能呢？传感器像人的眼睛、耳朵、鼻子等传感器官，是自动化生产线中的检测元件，能感受规定的被测量并按照一定的规律转换成电信号输出。

自动化生产线上常用的传感器有接近开关，位移测量传感器，压力测量传感器，流量测量传感器，温度、湿度检测传感器，成分检测传感器，图像检测传感器等许多类型。

3. 可编程控制器技术

自动化生产线的每一个单元都安装有一个可编程控制器（PLC）。它就像我们的大脑一样，思考每一个动作、每一招、每一式，指挥自动化生产线上的机械手按程序进行动作，是自动化生产线的核心部件。

4. 通信技术

在现代的自动化生产线中，不同的工作单元控制设备并非是独立运行，而是通过通信手段，相互之间进行信息交换，形成一个整体，从而提高了设备的控制能力、可靠性，实现了集中处理、分散控制。

5. 人机界面技术

人机交互设备提供的人机互动方式就像一个窗口,是操作者与 PLC 之间的对话窗口。人机界面以图形形式,显示 PLC 的状态、当前过程数据及故障信息。

 Part B

计算机集成制造系统(CIMS)

CIMS 是 computer integrated manufacturing system 的缩写。它是随着计算机辅助设计与制造的发展而产生的。CIMS 描述了一种制造、管理和公司运作的新方法。虽然 CIMS 包含许多先进的制造技术,如机器人、数控、计算机辅助设计、计算机辅助制造、计算机辅助工程和准时生产等技术,但它超越了这些技术。CIMS 是一种真正的柔性制造系统,不需要对系统做结构上的改变,就可以批量地制造各种各样的零件和部件。CIMS 包括完成全部过程所需的软件和自动化系统,包括产品设计、系统编程、生产成本估算、产品的实际制造、指令输入、库存跟踪及实际生产成本分析环节。

任务 5　机电设备的安全与维护技术

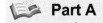 **Part A**

3.5.1　安　全　标　识

在机电设备上存在危险的地方通常有安全标识。我们必须理解每个安全标识的含义。常用安全标识见表 3-5-1。

表 3-5-1　常用安全标识

安全标识	安全标识
⚡ 高压危险	👷 注意防尘
⚠ 注意安全	♨ 当心高温
🔊 当心噪声	🚜 当心叉车

续表

安全标识	安全标识
 当心机械伤人	 当心吊物
 当心落物	 当心扎脚
 禁止合闸,有人工作	 禁止攀登,高压危险
 禁止操作,有人工作	 禁止靠近
 必须戴安全帽	 必须戴防护眼镜

3.5.2 劳动保护

劳动保护是国家和单位为保护劳动者在劳动生产过程中的安全和健康所采取的立法、组织和技术措施的总称。劳动保护的目的是为劳动者创造安全、卫生、舒适的劳动工作条件，消除和预防劳动生产过程中可能发生的伤亡、职业病和急性职业中毒，保障劳动者以健康的状态参加社会生产。常用防护措施如下：

1. 眼部防护

在生产车间里，采取眼部防护措施是在所有的安全规则中最为重要的一项。金属切屑或刨花能以很快的速度飞出，导致严重的眼部损伤。

2. 危害性噪声防护

在生产车间里，噪声危害是非常普遍的。高频噪声能引起永久性听力损伤。虽然噪声危害不可能被消除，但是通过戴减噪耳套、防噪声耳塞，或者同时佩戴，可以避免听力损伤。

3. 足部防护

在生产车间里，地面上经常被锋利的金属切屑覆盖，并且重型的原材料也可能砸到脚部。因此，安保鞋或坚固的皮鞋应该一直穿着。这些鞋在脚尖的部位都有一块抗压钢板。有些安保鞋还有脚背部防护。

电工作业时必须要穿戴合格的绝缘鞋。绝缘鞋作为从事电气工作的安全辅助用具，可以有效地降低触电的概率。

3.5.3 日常保养与维护

日常保养与维护是保养与维护工作中最重要的一种方法。如果你经常保养你的设备并且阅读保养手册，那么你就可以处理在工作中出现的很多小问题。为了使设备正常工作，保护保修权利，就必须遵守这些必要的规范。

1. 检查外观

① 每八小时班，应加满冷却液。
② 检查导轨润滑油箱液位。
③ 清理导轨防护罩和底板上的铁屑。
④ 清理换刀装置上的铁屑。
⑤ 检查齿轮箱中的油位。将油加到开始从机油箱底部的溢流管滴出为止。
⑥ 检查导轨防护罩是否正常运行，必要时用轻油润滑。
⑦ 检查所有软管和润滑管路是否破裂。
⑧ 检查润滑油过滤器，清除过滤器底部的残余物。

2. 检查控制单元内部

检查以下故障是否被排除：

① 电缆连接器松动。
② 电缆损伤。
③ 装配螺钉松动。
④ 固定放大器螺钉松动。
⑤ 冷却风扇工作异常。
⑥ 印制电路板插入不正确。

3.5.4 故障诊断

1. 故障诊断技术

故障诊断的内容包括确定故障类型和最有可能发生故障的位置、范围以及检测的时间。为了进一步提高设备的可靠性和安全性，早期自动诊断和故障定位是非常重要的。传统的途径是监测一些重要的变量，如温度、压力、振动，如果超过常规范围，即产生报警。

工艺技术操作要求越来越先进的管理和故障诊断，以提高可靠性、安全性和经济性。

2. 发生故障时的处理方法

当发生故障时，如何正确地把握发生了哪种类型的故障并采取相应的措施，对迅速修复设备是非常重要的。

按下述步骤检查故障（图 3-5-1）。

（1）何时发生的故障

① 故障发生的日期及时刻。

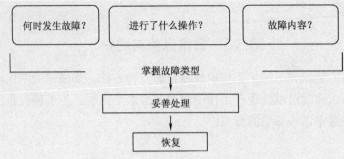

图 3-5-1　发生故障时的处理方法

② 操作时发生的吗？
③ 接通电源时发生的吗？
④ 是否在打雷、停电或对电源有干扰时发生的？发生了几次故障？
⑤ 只出现过一次吗？
⑥ 多次出现吗？（发生的频率，几次/时，几次/日，几次/月？）

（2）进行了何种操作后发生的故障
① 发生故障时系统是何种方式？
② 再次进行同样操作时是否发生同样故障？（确认故障再现性。）
③ 输入/输出数据时发生的故障吗？
④ 发生了与进给伺服轴有关的故障吗？
a. 低速进给、高速进给都发生故障吗？
b. 某一特定轴发生的故障吗？
⑤ 发生了与主轴有关的故障吗？
故障是在何时发生的？（接通电源时，加速时，减速时或通常运转时）

（3）发生了何种故障
① 报警显示画面上，显示什么报警内容？
② 画面显示是否正常？
③ 加工尺寸不准确吗？
a. 误差多大？
b. 位置显示画面是否正确？
c. 偏置量设定是否正确？

（4）其他信息
① 设备附近是否有干扰源？
② 设备内部是否采取了抗干扰的措施？
③ 检查输入电压：
a. 电压有无变动？
b. 有无相间电压差？
c. 是否供给标准电压？

④ 控制单元周围温度是多少？
⑤ 控制单元上是否有较大振动？

 Part B

颜 色 标 志

所有维修车间和工作场所，都应当标记上正确的颜色，用以区分危险、出口、安全通道和急救站。除了用正确、一致的颜色喷涂外，还可以常用如贴纸和纸带等办法。下面列出维修车间的主要标准颜色。

红色标志通常用于如下的设备和地点：

（1）火警箱

（2）灭火箱

（3）消防泵

（4）电力机械的急停按钮

（5）危险设备的急停门

（6）警告标志

白色背景的绿色标志通常用于如下的设备和地点：

（1）急救用的设备

（2）急救医务室

（3）机械的安全启动按钮

橙色标志通常用于机械设备或带电设备的危险部件。

Appendix A G-Codes and M-Codes for FANUC CNC

Table A-1 G-Codes

G-Codes（G 代码）	Group（组）	Function（功能）
G00*	01	rapid positioning（快速定位）
G01		linear interpolation（直线插补）
G02		clockwise circular interpolation, CW（顺时针圆弧插补）
G03		counterclockwise circular interpolation, CCW（逆时针圆弧插补）
G04	00	dwell（暂停）
G10		programmable data input（可编程数据输入）
G11		programmable data input cancel（取消可编程数据输入）
G15*	17	polar coordinates command cancel（取消极坐标指令）
G16		polar coordinates command（极坐标指令）
G17*	02	X-Y plane selection（X-Y 平面选择）
G18		Z-X plane selection（Z-X 平面选择）
G19		Y-Z plane selection（Y-Z 平面选择）
G20*	06	inch format［英制(in)］
G21		metric format［米制(mm)］
G27	00	return reference point check（参考点返回检查）
G28		automatic return to reference point（自动返回参考点）
G29		return from reference point（从参考点返回）
G40*	07	cutter radius compensation cancel（取消刀具半径补偿）
G41		cutter radius compensation-left（刀具半径左补偿）
G42		cutter radius compensation-right（刀具半径右补偿）
G43	08	tool length compensation（刀具长度补偿）
G49*		tool length compensation cancel（取消刀具长度补偿）
G50	11	coordinate system setting and maximum rpm（设定工件坐标系或最大主轴转速）
G50.1*	22	programmable mirror image cancel（取消可编程镜像）
G51.1		programmable mirror image（可编程镜像有效）
G54~G59	12	select workpiece coordinate system（选择工件坐标系）
G65	00	custom macro simple call（宏程序调用）
G66	12	custom macro modal call（宏程序模态调用）
G67*		custom macro modal call cancel（取消宏程序模态调用）

Appendix A G-Codes and M-Codes for FANUC CNC

(continued)

G-Codes（G 代码）	Group（组）	Function（功能）
G73	09	high-speed peck drilling cycle（高速深孔钻孔循环）
G74		counter tapping cycle（反攻螺纹循环）
G76		fine boring cycle（精镗循环）
G80*		canned cycle cancel（钻孔固定循环取消）
G81		drilling cycle(feed in, rapid out)［钻孔循环（进刀，快速退刀）］
G84		tapping cycle(feed in, feed out)［攻螺纹循环（进刀，退刀）］
G85		boring cycle(feed in, feed out)［镗孔循环（进刀，退刀）］
G86		boring cycle(feed in, stop-wait, rapid out)［镗孔循环（进刀，暂停，快速退刀）］
G87		back boring cycle(peck, rapid out)［背镗循环（慢速进刀，快速退刀）］
G89		boring cycle(feed in, dwell, feed out)［镗孔循环（进刀，暂停，退刀）］
G90*	03	absolute programming（绝对坐标编程）
G91		incremental programming（增量坐标编程）
G92	00	preset part program zero point（工件坐标原点设置）
G94*	05	feed per minute mode（每分进给方式）
G95		feed per rotation（每转进给）
G98*	10	initial point return（返回初始点平面）
G99		R plane return（返回 R 点平面）

Note: *expressed as initial stale.
注：*表示初始状态。

Table A-2 M-Codes

M-Codes（M 代码）	Type（类型）	Function（功能）
M00	B	program stop(non-modal)［程序停止（非模态）］
M01		optional stop(non-modal)［程序选择停止（非模态）］
M02		end of program(non-modal)［程序结束（非模态）］
M03	A	spindle CW（主轴顺时针旋转）
M04		spindle CCW（主轴逆时针旋转）
M05	B	spindle stop（主轴停止）
M06		tool change(non-modal)［自动换刀（非模态）］
M07	A	flood coolant on（冷却液打开）
M08		mist coolant on（喷雾冷却液打开）
M09	B	coolant off（冷却液关闭）
M30		end of tape（程序结束并返回）
M97	A	local subprogram call（部子程序调用）
M98		subprogram call（子程序调用）
M99		subprogram end (return to main program)［子程序调用结束（返回主程序）］

Note:

A: Those executed with the start of axes movement in a block.

B: Those executed after the completion of axes movement in a block.

注：

A：在轴运动前被执行。

B：在轴运动完成后被执行。

Table A-3　Letters

Letters（字母）	Function（功能）	
A	rotary axes/旋转轴	angular coordinate around X-axis（绕X轴的旋转坐标）
B		angular coordinate around Y-axis（绕Y轴的旋转坐标）
C		angular coordinate around Z-axis（绕Z轴的旋转坐标）
D	tool diameter offset designation（指定刀具的直径偏置）	
F	linear feed rate（进给速度）	
G	preparatory code［准备代码（G代码）］	
H	tool height offset designation（指定刀具的高度偏置）	
I	axes used as auxiliary of X,Y,Z axes / X,Y,Z轴的辅助尺寸	incremental distance to arc center along X-axis from start（起点到圆弧中心在X轴方向的相对距离）
J		incremental distance to arc center along Y-axis from start（起点到圆弧中心在Y轴方向的相对距离）
K		incremental distance to arc center along Z-axis from start（起点到圆弧中心在Z轴方向的相对距离）
L	number of times to loop a canned cycle/固定循环的重复次数	
M	miscellaneous code［辅助代码（M代码）］	
N	line number（行号）	
O	program name（程序名）	
P	dwell time（暂停时间）	
Q	peck depth（步进切入深度）	
R	Z-coordinate of retract plane（退刀平面的Z坐标）	
S	spindle speed(r/min)［主转速（r/min）］	
T	tool number designation（指定刀具号）	
X	linear axes/线性轴	coordinate on X-axis（X坐标轴）
Y		coordinate on Y-axis（Y坐标轴）
Z		coordinate on Z-axis（Z坐标轴）
U	secondary linear axes/第二线性坐标轴	parallel to X-axis（平行于X坐标轴）
V		parallel to Y-axis（平行于Y坐标轴）
W		parallel to Z-axis（平行于Z坐标轴）

Appendix B Abbreviations

Table B-1 Abbreviations

缩写	全写	中文释义
3-D	three dimension	三维，三坐标，三轴
APC	automatic pallet changer	自动托盘交换装置
APL	automatic production line	自动化生产线
APT	automatically programmed tool	自动编程系统
ATC	automatic tool changer	自动换刀装置
AWF	automatic wire feed	自动走丝
CAD	computer-aided design	计算机辅助设计
CADD	computer-aided drafting and design	计算机辅助绘图与设计
CAE	computer-aided engineering	计算机辅助工程
CAI	computer-aided instruction	计算机辅助教学
CAIN	computer-aided inspection	计算机辅助检测
CAM	computer-aided manufacture	计算机辅助制造
CAN	cancel	取消
CAP	computer-aided programming	计算机辅助编程
CAPP	computer-aided process planning	计算机辅助工艺规程设计
CAQC	computer-aided quality control	计算机辅助质量控制
CCW	counter clockwise	逆时针旋转，反转
CIMS	computer integrated manufacturing system	计算机集成制造系统
CIPM	computer integrated production management	计算机集成生产管理
CNC	computer numerical control	计算机数字控制
CPU	central processing unit	中央处理单元
CRT	cathode ray tube	阴极射线管
CW	clockwise	顺时针旋转，正转
DGNOS	diagnostic	诊断
DIR	directory	目录
DNC	direct numerical control	直接数字控制
DRN	dry run	空运行
DRO	digital readout	数显装置
EDM	electricity discharge machine	电火花加工机床

(continued)

缩 写	全 写	中文释义
EOB	end of block	程序段结束
EOF	end of file	文件结束
EPROM	erasable programmable read only memory	可擦除可编程只读存储器
E-stop	emergency stop	紧急停止
FA	factory automation	工厂自动化
FBD	function block diagram	功能块图
Fig.	figure	图
FMC	flexible manufacturing cell	柔性制造单元
FMS	flexible manufacturing system	柔性制造系统
GT	group technology	成组技术
HMC	horizontal machining center	卧式加工中心
HP	horse power	马力，功率
HSS	high-speed steel	高速钢
ID	inside diameter	内径
INC	increment	增量
ISO	international standard organization	国际标准化组织
LAD	ladder diagram	梯形图
LCS	local coordinate system	局部坐标系
MCP	manual control panel	手动控制面板
MCS	machine coordinate system	机床坐标系
MCU	machine control unit	机床控制单元
MDI	manual data input	手动数据输入
MLK	machine lock	机床锁住
MPG	manual pulse generator	手摇脉冲发生器
MPP	manufacturing process planning	制造工艺规程
MTB	machine tool builder	机床制造商
OD	outside diameter	外径
OSP	option stop	选择停止
PLC	programmable logical control	可编程控制器
POSIT	position	位置
PROG	program	程序
PTP	point-to-point	点到点，点位
RAM	random access memory	随机存储器
REF	reference point	回参考点
RET	return	返回
ROM	read-only memory	只读存储器

Appendix B Abbreviations

(continued)

缩　　写	全　　写	中 文 释 义
RP	rapid prototyping	快速原型技术
RPM	revolutions per minute	转每分
SBK	single block	单段
SC	servo control	伺服控制系统
SFC	sequential function chart	顺序功能图
STL	statement table	语句表
SV	servo	伺服
VMC	vertical machining center	立式加工中心
WCS	workpiece coordinate system	工件坐标系
WEDM	wire cut electrical discharge machining	电火花线切割加工
WEDM-HS	high-speed WEDM	高速（快）走丝线切割机床
WEDM-LS	low-speed WEDM	低速（慢）走丝线切割机床
ZRN	zero return	回零

References

[1]沈言锦．电气自动化专业英语[M]．北京：机械工业出版社，2016．

[2]康辉，杨晓辉．电气专业英语[M]．北京：机械工业出版社，2016．

[3]李桂云．数控技术应用专业英语[M]．3版．大连：大连理工大学出版社，2014．

[4]吕栋腾．机电英语[M]．北京：北京邮电大学出版社，2014．

[5]徐存善．机械电气专业英语[M]．北京：机械工业出版社，2011．

[6]吕景泉．自动化生产线安装与调试[M]．2版．北京：中国铁道出版社，2009．

[7]黄星，夏玉波．机电一体化专业英语[M]．北京：人民邮电出版社，2010．

[8]别传爽．机电专业英语[M]．北京：北京理工大学出版社，2010．

[9]赵金广．数控技术英语[M]．北京：北京理工大学出版社，2011．

[10]杨桂慧，孙亮波．机械专业英语[M]．北京：国防工业出版社，2010．

[11]杨宏，邱文萍，洪霞．机电与数控专业英语[M]．武汉：武汉大学出版社，2009．

[12]汤彩萍．机械专业交际英语[M]．北京：电子工业出版社，2011．

[13]王兆奇，刘向红．数控专业英语[M]．2版．北京：机械工业出版社，2012．

[14]鲍海龙．数控专业英语[M]．北京：机械工业出版社，2011．

[15]沈言锦．机电专业英语[M]．北京：机械工业出版社，2011．